BIOLOGICAL SYSTEMS in MARINE LIFE

BIOLOGICAL SYSTEMS
in
MARINE LIFE

Dr. Shweta Sharma

Oxford Book Company
Jaipur ▪ New Delhi

Published By

Oxford Book Company

Regd. Off. "Bony Residency", Gate No. 2
Opp. Tilak Public School, Vishweswariya Nagar,
Gopalpura Road, Jaipur - 302018, India
Ph: 0141 - 2761280, Telefax: 2761381
email - oxfordbookjpr@gmail.com
website - oxfordbookcompany.com

Delhi Office: 102, 1st Floor, Satyam House,
4327/3, Ansari Road, Darya Ganj,
New Delhi - 110002, India
Ph: 011-45652440

Distributed By

KEYA INTERNATIONAL PRESS LLP
E-mail: keyainternationalpressllp@gmail.com

Published in India, 2025

ISBN : 9789391096618

Typeset by Shivangi Computers, Jaipur

Printed and bound in India by Thomson Press Ltd.,
New Delhi

Contents

Preface

The study of marine biology includes a wide variety of disciplines such as astronomy, biological oceanography, cellular biology, chemistry, ecology, geology, meteorology, molecular biology, physical oceanography and zoology and the new science of marine conservation biology draws on many longstanding scientific disciplines such as marine ecology, biogeography, zoology, botany, genetics, fisheries biology, anthropology, economics and law. Like all scientific disciplines, the study of marine biology also follows the scientific method. The overriding goal in all of science is to find the truth. Although following the scientific method is not by any means a rigid process, research is usually conducted systematically and logically to narrow the inevi margin of error that exists in any scientific study, and to avoid as much bias on behalf of the researcher as possible. The primary component of scientific research is characterisation by observations. Hypotheses are then formulated and then tested based on a number of observations in order to determine the degree to which the hypothesis is a true statement and whether or not it can be accepted or rejected. Testing is then often done by experiments if hypotheses can produce predictions based on the initial observations.

The behaviour of marine animals is studied so that we understand the animals that share the planet with us. This is also an important field for help in understanding how to protect endangered species, or how to help species whose habitats are threatened by man or natural phenomena. The study of marine animal behaviour usually falls under the category of ethology because most often marine species must be observed in their natural environment, although there are many marine species observed in controlled environments as well. Sharks are most often studied in their natural habitat for obvious reasons. Life in the sea has been a subject of fascination for thousands of years. One of the most important reasons for the study of sea life is simply to understand the world in which we live. The oceans cover 71% of this world, and yet we have only scratched the surface when it comes to understanding them. Scientists estimate that no more than 5% of the oceans have been explored.

The difference between the terms "marine biology" and "biological oceanography" is subtle, and the two are often used interchangeably. As mentioned above, marine biology is the study of marine species that live in the ocean and other salt-water environments. Biological oceanography also studies marine species, but in the context of oceanography. So a biological oceanographer might study the impact of cold upwellings on anchovy populations off the coast of South America, where a marine biologist might study the reproductive behaviour of anchovies. The book is the first reference book of its kind dealing with a broad spectrum of Microbiology of Marine Life and environmental pollution problems.

—Dr. Shweta Sharma

Chapter 1

Introduction

Simply put, marine biology is the study of life in the oceans and other saltwater environments such as estuaries and wetlands. All plant and animal life forms are included from the microscopic picoplankton all the way to the majestic blue whale, the largest creature in the sea—and for that matter in the world.

The study of marine biology includes a wide variety of disciplines such as astronomy, biological oceanography, cellular biology, chemistry, ecology, geology, meteorology, molecular biology, physical oceanography and zoology and the new science of marine conservation biology draws on many longstanding scientific disciplines such as marine ecology, biogeography, zoology, botany, genetics, fisheries biology, anthropology, economics and law.

Like all scientific disciplines, the study of marine biology also follows the scientific method. The overriding goal in all of science is to find the truth. Although following the scientific method is not by any means a rigid process, research is usually conducted systematically and logically to narrow the inevi margin of error that exists in any scientific study, and to avoid as much bias on behalf of the researcher as possible. The primary component of scientific research is characterisation by observations. Hypotheses are then formulated and then tested based on a number of observations in order to determine the degree to which the hypothesis is a true statement and whether or not it can be accepted or rejected. Testing is then often done by experiments if hypotheses can produce predictions based on the initial observations.

The essential elements of the scientific method are iterations and recursions of the following four steps:

- Characterisation (observation)
- Hypothesis (a theoretical, hypothetical explanation)

- Prediction (logical deduction from the hypothesis)
- Experiment (test of all of the above).

These steps are all used in the study of marine biology, which includes numerous sub fields including:

Microbiology: The study of microorganisms, such as bacteria, viruses, protozoa and algae, is conducted for numerous reasons. One example is to understand what role microorganisms play in marine ecosystems. For example, bacteria are critical to the biological processes of the ocean, as they comprise 98% of the ocean's biomass, which is the total weight of all organisms in a given volume. Microbiology is also important to our understanding of the food chain that connects plants to herbivorous and carnivorous animals. The first level in the food chain is primary production, which occurs at the microbial level. This is an important biological activity to understand as primary production drives the entire food chain.

Scientists also study marine microbiology to find new organisms that may be used to help develop medicines and find cures for diseases and other health problems.

Fisheries and Aquaculture: to protect biodiversity and to create sustainable seafood sources because of the world's dependence on fish for protein. There are many areas of study in this field.

The ecology of fisheries includes the study of their population dynamics, reproduction, behaviour, food webs, and habitat.

Fisheries management includes studies on the impact of overfishing, habitat destruction, pollution and toxin levels, and ways to increase populations for sustainability as seafood.

Aquaculture includes research on the development of individual organisms and their environment. The objective is most often to develop the knowledge needed to cultivate certain species in a designated area in open water or in captivity in order to meet consumer demand. Technological advances have enabled seafood "farms" to produce high-demand products that traditional commercial fisheries cannot meet. This is a controversial area however, and an issue that will become of greater importance as our fish stocks continue to decline.

Environmental Marine Biology: includes the study of ocean health. It is important for scientists to determine the quality of the marine environment to ensure water quality is sufficient to sustain a healthy environment. Coastal environmental health is an important area of environmental marine biology so that scientists can determine the impact of coastal development on water quality for the safety of people visiting the beaches and to maintain a healthy marine environment. Pollutants, sediment, and runoff are all potential threats to marine health in coastal areas. Offshore marine environmental health is also studied.

For example, an environmental biologist might be required to study the impact of an oil spill or other chemical hazard in the ocean. Environmental biologists also study Benthic environments on the ocean bottom in order to understand such issues as the chemical makeup of sediment, impact of erosion, and the impact of dredging ocean bottoms on the marine environment.

Deep-sea Ecology: advances in technology of equipment needed to explore the deep sea have opened the door to the study of this largely unknown space in the sea. The biological characteristics and processes in the deep-sea environment are of great interest to scientists. Research includes the study of deep ocean gases as an alternate energy source, how animals of the deep live in the dark, cold, high pressure environment, deep sea hydrothermal vents and the lush biological communities they support.

Ichthyology: is the study of fishes, both salt and freshwater species. There are some 25,000+ species of fishes including: bony fishes, cartilaginous fishes, sharks, skates, rays, and jawless fishes. Ichthyologists study all aspects of fish from their classification, to their morphology, evolution, behaviour, diversity, and ecology. Many ichthyologists are also involved in the field of aquaculture and fisheries.

Marine Mammology: This is the field of interest to most aspiring marine biologists. It is the study of cetaceans families of whales and dolphins, and pinnipeds (seals, sea lions, and the walrus). Their behaviours, habitats, health, reproduction, and populations are all studied. These are some of the most fascinating creatures in the sea; therefore, this is an extremely competitive field, and difficult to break into because the competition for research funding is also quite heavy.

One area of research currently being conducted on whales is the impact of military sonar on their health and well-being. The scientific community believes that high frequency sound waves cause internal damage and bleeding in the brains of whales, yet the military denies this claim. Military sonar can also interfere with the animal's own use of sonar for communication and echolocation. More research is needed; however, in recent years science has proven the claims to be valid and the military has begun limiting its use of sonar in specific areas.

Marine Ethology: The behaviour of marine animals is studied so that we understand the animals that share the planet with us. This is also an important field for help in understanding how to protect endangered species, or how to help species whose habitats are threatened by man or natural phenomena. The study of marine animal behaviour usually falls under the category of ethology because most often marine species must be observed in their natural environment, although there are many marine species observed in controlled environments as well. Sharks are most often studied in their natural habitat for obvious reasons.

Why Study Marine Biology?

Life in the sea has been a subject of fascination for thousands of years. One of the most important reasons for the study of sea life is simply to understand the world in which we live. The oceans cover 71% (and rising) of this world, and yet we have only scratched the surface when it comes to understanding them. Scientists estimate that no more than 5% of the oceans have been explored. Yet, we need to understand the marine environment that helps support life on this planet, for example:

How is Marine Biology Studied?

Advances in technology have opened up the ocean to exploration from the shallows to the deep sea. New tools for marine research are being added to the list of tools that have been used for decades such as:

Trawling:- has been used in the past to collect marine specimens for study, except that trawling can be very damaging to delicate marine environments and it is difficult to collect samples discriminately. However when used in the midwater environment, trawls can be every effective at collecting samples of elusive species with a wide migratory range.

Plankton Nets : plankton nets have a very fine weave to catch microscopic organisms in seawater for study.

Remotely Operated Vehicles (ROVs) : have been used underwater since the 1950s. ROVs are basically unmanned submarine robots with umbilical cables used to transmit data between the vehicle and researcher for remote operation in areas where diving is constrained by health or other hazards. ROVs are often fitted with video and still cameras as well as with mechanical tools for specimen retrieval and measurements.

Underwater Habitats : the National Oceanic and Atmospheric Administration (NOAA) operates Aquarius , a habitat 20 m beneath the surface where researchers can live and work underwater for extended periods.

Fibre Optics : Fiber optic observational equipment uses LED light (red light illumination) and low light cameras that do not disturb deep-sea life to capture the behaviours and characteristics of these creatures in their natural habitat.

Satellites : are used to measure vast geographic ocean data such as the temperature and colour of the ocean. Temperature data can provide information on a variety of ocean characteristics such as currents, cold upwelling, climate, and warm water currents such as the Gulf Stream. Satellites are also used for mapping marine areas such as coral reefs and for tracking marine life tagged with sensors to determine migratory patterns.

Sounding : hydrophones, the microphone's counterpart, detect and record acoustic signals in the ocean. Sound data can be used to monitor waves, marine mammals, ships, and other ocean activities.

Sonar : similar to sounding, sonar is used to find large objects in the water and to measure the ocean's depth (bathymetry). Sound waves last longer in water than in air, and are therefore useful to detect underwater echoes.

Computers : sophisticated computer technology is used to collect, process, analyse, and display data from sensors placed in the marine environment to measure temperature, depth, navigation, salinity, and meteorological data. NOAA implemented computer technology aboard its research vessels to standardise the way this data is managed.

Marine Biology versus Biological Oceanography

The difference between the terms "marine biology" and "biological oceanography" is subtle, and the two are often used interchangeably. As mentioned above, marine biology is the study of marine species that live in the ocean and other salt-water environments. Biological oceanography also studies marine species, but in the context of oceanography. So a biological oceanographer might study the impact of cold upwellings on anchovy populations off the coast of South America, where a marine biologist might study the reproductive behaviour of anchovies.

Division of Marine Life Science

The division is divided into six chairs: Aquaculture Biology, Aquaculture Genetics and Genomics, Marine Biotechnology and Microbiology, Marine Bioresources Chemistry, Marine Products and Food Sciences, Marine Biosafty Science and Technology (fixed term contract).

These chairs study 18 fields, and their activities cover a wide research area of marine biology and chemistry, especially comparative and reproductive physiology, developmental and breeding biology, bimolecular and biopolymer chemistry, marine microbiology, analytical and functional biochemistry, biochemical-process technology and food science and technology. The division is well equipped with an experimental aquarium, training factory for the utilisation of marine bioresources, an electron microscope, gas chromatograph-mass spectrometer, high-performance liquid chromatograph-mass spectrometer, and a nuclear magnetic resonance spectrometer.

Aquaculture Biology

As a fundamental study to utilise aquatic animals as bioresources, this laboratory deals with physiological and biochemical events using various techniques extending from histological to moleculer levels. Three major fields are comparative physiology, reproductive physiology, and functional biochemistry.

Comparative Physiology

Major research subjects in this laboratory are; 1) reproduction, metabolism and growth of aquatic invertebrates, and 2) hard tissue physiology of fish and aquatic invertebrates. Through the physiological, histological and molecular research, researchers and technicians who specialise animal physiology, not only in the field of fisheries science but also in the field of general biology, are fostered.

Reproductive Physiology

We carry out the advanced education and research on the molecular mechanisms of gonadal sex differentiation, oogenesis and spermatogenesis in mainly fish using physiological, histological, molecular endocrinological, biochemical and cell culture techniques. The objective is to establish techniques for seed production in various species of marine animals, and to develop well-qualified persons. We also carry out the education and research on the development of environment-friendly closed aquaculture system.

Functional Physiology

We are studying precursor proteins of egg yolk proteins (vitellogenin) and envelop proteins (choriogenin) during vitellogenesis and choriogenesis is salmonid species using immunobiochemical technique. The objective is to understand the mechanism of the protein synthesis, secretion and accumulation after establishment of identification, purification and measurement of the proteins.

Aquaculture Genetics and Genomics

The advanced education of graduate students majoring in Aquatic Plant Genetics, Aquatic Animal Genetics and Aquatic Bioresource Genetics, with particular reference to; genetics and genome science for theoretical basis of breeding and stock improvement of aquatic animals and benthic marine algae; biotechnology for methodological promotion of aquaculture; the conservation, management and sustainable use of aquatic genetic resources for breeding.

Plant Aquaculture Genetics

Applying our basic knowledge of developmental biology, physiology, genetics and molecular biology of aquatic plants, the development of model aquatic plants and genome analysis, and breeding, multiplication and utilisation of seaweed resources are actively performed.

Animal Aquaculture Genetics

As the theoretical and technological base of aquaculture and genetic improvement of commercially important fish and aquatic animals, gene expression, atypical reproduction (unreduced gametogenesis, unisexuality, natural gynogenesis, etc.), chromosome manipulation, embryo manipulation and comparative genomics using molecular markers are actively studied and educated.

Aquatic Bioresource Genetics

Genetic characterisation of wild and cultured populations of marine vertebrates and invertebrates; genome analysis for development of polymorphic molecular markers; chromosome mapping of DNA markers for construction of cytogenetic map; molecular evolution of fish genome; molecular population genetics for conservation, management and sustainable use of aquatic organisms.

Marine Biotechnology and Microbiology

We are in charge of graduate school education on microbiology, molecular biology and epizootiology for aquatic organisms. One of our goal is to advance the sciences regarding comprehensive physiology and ecology for aquatic organisms by molecular and cell biological techniques. Another goal is to prevent and control infectious diseases of cultured organisms.

Marine Microbiology

In this field, main subjects are taxonomy, ecology and host-microbes interaction of marine microbes by using physiological, biochemical and molecular biochemical techniques. We are also searching for useful gene resources of marine bacteria as subject of applied Microbiology.

Main researching themes: Identification and classification of microbe of disease associate with marine animals and plants. Searching for symbiotic marine bacteria associated with marine animals. Molecular ecology of pathogenic bacteria and symbiotic bacteria associated with marine animals. Searching for sensing factors possessing pathogenic bacteria and symbiotic bacteria associated with marine animals to host animals.

Searching for valuable enzyme-coding genes from marine environmental DNAs.

Marine Molecular Biology

Laboratory of Maine Molecular Biology deals with genetic engineering for various functional proteins and enzymes from marine animals and plants, and their applications to industrial uses.

Current Research Subjects are as Follows

Isolation and cDNA cloning of industrially useful enzymes, cellulose, alginate lyase, mannanase, protease, from marine invertebrates such as scallop, abalone, and sea urchin.

Mass production of recombinant enzymes of marine organisms by using bacterial and insect cells. Production of protoplasts from brown seaweeds by using polysaccharide-degrading enzymes from marine invertebrates.

Isolation and characterisation of enzymes related to growth and differentiation of brown seaweeds. Study on regulatory mechanism of troponin from scallop adductor muscles.

Aquatic Epizootiology

Outbreaks of diseases are one of the most serious problems on aquacultured organisms for securing of food source. In our laboratory, we focus on infectious diseases of fish and shellfish, especially diagnosis and treatment of the diseases, moreover molecular and microbiological studies on the causative agents of the diseases. Also, we perform studies on prevention of diseases for aquatic organisms.

Marine Bioresource Chemistry

The laboratory consisted of two fields, Marie Biofunctional Chemistry and Marine Bio-analytical Chemistry, is engaged in the education for the graduate school.

The lectures on the understanding of unique life in marine organism, characterisation of unique compounds for marine life, functionalities of newly found compounds from marine organism, and maximal utilisation of marine resources as food, nutraceutical, and pharmaceutical compounds.

Biofunctional Chemistry

Marine resources contain a variety of functional compounds. We study on the chemical properties, nutritional effects, biological functions, health benefits, and utilisation of these marine bioresources, and their application to food materials, nutraceuticals and medicines. Absorption, metabolism, nutrigenomics, and molecular biology of marine functional compounds are also our main research themes.

Our research project also includes stabilisation of functional components in food systems and their enzymatic modification for higher biological activities.

Bio-analytical Chemistry

This field has focused on teaching and researching on analytical chemistry stereochemistry and biochemistry of marine organic compounds, particular lipids and lipid related compounds.

Current research activities: stereochemistry of glycerolipids in fish, algae and bacteria; stereospecificity of lipid related enzymes such as lipases; precision analysis of fish oils; analysis, distribution and function of marine organic compounds with unusual structures.

Biomolacular Chemistry

The field is engaged in the education and research on the understanding of unique properties of various compounds in marine resources, and the development of the purification, analysis separation, isolation, and evaluation systems.

Marine Products and Food Science

This laboratory provides research training for graduate students and also carries out various other activities which include development of new theories and highly improved technologies for effective utilisation of marine bioresources as foods and other value added or highly consumed products.

Graduate students are trained in specialised fields and research along their priorities of interest in interdisciplinary fields including chemistry, biology, biochemistry, microbiology, and also production engineering in the related fields.

Marine Bioresources Utilisation

To explore the useful potentials as health beneficial, drugs and for other purposes materials from marine bioresources are the mission of this field. We are identifying some beneficial enzyme inhibitors such as against glucosidase, urease, glucuronidase and so forth from algae. Our current topic is the discovery of a beneficial therapeutic potential against tumour of marine phospholipid liposomes which act as an immune system booster in combination with glucan.

Food Biochemistry

With a view to utilise the fish and shellfish at a high-degree, we educate and study on the biochemical properties of muscle proteins and endogenous enzymes, such as proteinases and transglutaminases, which are closely related to the freshness and quality of seafood. We also work on the structure-activity relationship of fish and shellfish proteins by using molecular biological and genetic techniques. Based on these studies, we attempt to develop a novel preservation methods and to make new seafood products by control of the endogenous enzymes in fish and shellfish.

Food Functional Chemistry

This field covers the applied science and technology for high utilisation of marine bioresource proteins as food materials and bioproducts. In recent topics are as follows: (1) Applied protein chemistry: Attempts to improve protein-functionality (food process characteristics and biofunctions) using molecular modification. (2) Seafood allergy: Identifying allergenic proteins in seafood and effect of food processing on the allergenicity. (3) Applied marine enzymology: Search for new enzymes and enzyme-inhibitors from unutilised marine bioresources and developing their application.

Food Safety

This field includes food preservation and food hygiene, and deals with following subjects to assure wholesomeness of marine foods: (1) specification of causative microorganisms and chemical substances of food deterioration and food-borne diseases, (2) analyses of mode of the actions and development of controlling methods with , cellular biological and molecular biological techniques, and (3) classification and beneficial utilisation of food microorganisms.

Chair of Marine Biosafety Science and Technology

The total safety of marine organisms from the production process to consumption is an urgent social need. This laboratory is designed to educate students and conduct research on both Marine Environmental Biosafety Management and the Marine Life Safety and Quality Management of Food.

Marine Environmental Biosafety Management

The main objectives of this field are the management of aquaculture environment and production of safe aquaculture feed from fisheries wastes for sustainable aquaculture. We provide education and conduct research on the separation science and technology for the removal of toxic substances from aquatic environments and feed ingredients.

Food Chain Safety and Quality Management

To ensure the safety and reliability on food chain, we have to continue the analysis and control of biological, chemical and physical hazard factors during growth, harvesting, processing, distribution and consumption of food. We try to systematize the safety management of food chain. The other target of this chair is to train persons who have highly expert knowledge about international assurance system of food.

Microbiology

Marine microbiology is the study of marine life forms that cannot be seen with the naked eye, including viruses, bacteria, fungi, protozoa and microscopic algae. Representatives of each group have been found in almost every marine habitat examined, from tropical coral reefs to the Antarctic and the greatest depths of the ocean. Microbes occur in the water column and sediments, on the surfaces of marine plants, animals and inanimate objects, in the intestinal tracts of fishes and invertebrates and even within the tissues of other living things. There is a staggering diversity of forms and modes of life among marine microbes. Some require light for growth, many require the saline nature of sea water, while others can survive without oxygen.

Because of their small size, the importance of microbes to life in the oceans is often not appreciated. Knowledge of their activities assumes a new importance when one considers that each millilitre of sea water may be the home for millions of individuals, and each gram of sediment may support thousands of millions of microbes. Microbes play a crucial role in ocean food webs, which support the diversity of life in the sea. Scientists therefore study the microscopic plants that fix carbon dioxide, the bacteria which grow on the myriad of organic, inorganic, dissolved and particulate chemicals in the sea, and the protozoa which feed on them. In this way knowledge is gained about how each life form contributes to all levels of the food web.

Other microbiologists investigate the unique ways in which microbes change chemicals from one form to another, and the importance of this to the health of the oceans. For example, particular kinds of microbes carry out some steps in the cycling of nitrogen from a dissolved gas in the sea to proteins in plants, animals and microbes, to inorganic forms and finally back to a gas. Maintenance of this cycle is essential for life in the oceans.

Many aspects of marine microbiological research have the potential to provide practical information and to form the basis of new biotechnological developments. Microbiologists studying diseases of marine organisms are providing disease diagnosis and control strategies to those attempting to grow marine animals such as prawns, molluscs and fish. New chemicals with potential industrial applications, such as friction-reducing lubricants on the surfaces of some fish, have been found to be produced by marine microbes. Major efforts are also being put into the search for new pharmaceuticals from marine microbes, such as antibiotics and anticancer agents.

The incredible diversity of microbes, their wide distribution in marine habitats, metabolic capabilities and intimate involvement in processes occurring in the oceans ensures lifetimes of discovery for scientists. Research into the activities of microbes involves opportunities for collaboration with scientists from disparate disciplines such as botany, geology, zoology, chemistry and marine biology. Field work and laboratory analysis using traditional microbiological methods as well as sophisticated techniques such as DNA technology are all tools of trade for the marine microbiologist. Marine microbiologists require tertiary qualifications and are employed by universities, a variety of state and federal departments and instrumentalities, and by private industry.

Advances in Biological Science Raise Troubling Questions about What it Means to be Human

By the end of the 21st century," writes Reason magazine science editor Ronald Bailey in his book "Liberation Biology," "the typical American may attend a family reunion in which five generations are playing together. And great-great-great grandma, at 150 years old, will be as vital as her 30 year old great-great grandson with whom she's playing touch football."

UCLA futurist Gregory Stock predicts in "Redesigning Humans" that the genetic engineering of progeny for health, intelligence, physical beauty, even sociability, will be so successful that procreation through intercourse will be deemed "too unpredictable," making "laboratory conception obligatory rather than optional."

Princeton biologist Lee Silver believes fervently, as described in "Remaking Eden," that the wonders of human redesign will eventually lead to a "special point" where our posterity will create themselves into a "special group of mental beings who "are as different from humans as humans are from primitive worms. 'Intelligence' will "not do justice to their cognitive abilities. 'Knowledge' does not explain the depth of their understanding. 'Power' is not strong enough to describe the control they have over technologies that can be used to shape the universe in which they live."

The prospect of a 150 year old living human being sounds fantastical. So does predesigning children or future generations with godlike powers. But many futurists and scientists say we humans are about to seize control of our own evolution.

If the impeders of scientific progress can just be pushed out of the way, they predict, the wonders of science and biotechnology will recreate us into superior beings who will live longer, look better, play harder and think smarter than any of us can even dream of doing today.

Others (including this writer) such scenarios as more hype than hope.

Some of us also worry that advocates of unfettered research are changing science from a means into an end, a belief system rather than a method.

Indeed, "bioskeptics," as they are sometimes called, see a utopian ideology of "scientism" forming that threatens to upend society's belief in human equality and unleash a "new eugenics," in which Aldous Huxley's dystopian vision of mind control from "Brave New World" could become a reality.

Look out America: The trajectory of science is coming into conflict with venerable human values. Which side prevails will depend less on what scientists can do than upon the ethical principles that govern society in an era of biological control.

In the United States today, every human being who is born possesses full moral and legal rights. But this is under pronounced assault. Influential philosophers, such as Princeton University's Peter Singer, assert that basing an individual's moral value on humanhood is irrational ain place of humanness as the criterion for ultimate value, these advocates offer "personhood theory," in which rights belong to "persons," a status earned by any organism or machine possessing minimal cognitive capacities.

If personhood theory ever governs society, the impact would be incalculable, for as futurist James Hughes writes in "Citizen Cyborg," "Persons don't have to be human, and not all humans are persons."

Opponents of personhood theory warn that it would lead to the most vulnerable humans being exploited as mere objects.

They note that some supporters of personhood theory already advocate infanticide for profoundly disabled babies and organ harvesting from people diagnosed to be in a persistent vegetative state.

The already simmering humanhood versus personhood controversy is going to boil over as our scientific and biotechnological capacities advance.

For example, what if it becomes technologically feasible to create cloned human embryos and gestate them in real or artificial wombs to fetal stage for use in drug testing or for organ procurement? (Such experiments have already been conducted successfully in cows.)

Those who believe that humanhood provides intrinsic value argue that such "fetal farming" should be prohibited because it reduces nascent human life to the status of a mere harvestable commodity.

Personhood theorists, on the other hand, would tend to support using cloned fetal nonpersons to save the lives of persons and to reduce the suffering of animals currently used in medical research, which are seen as having greater moral value because they possess higher cognitive capacities.

Promoters of this research point out that great good could result, for example, from obtaining human proteins from the milk of these altered animals for use in pharmaceuticals, a process known as "pharming." Other than a scattering of environmentalists and animal rights activists, few object to creation of these "transgenic" animals.

But important questions remain: How much human DNA in an animal is too much? (James Hughes believes that chimps should be "uplifted," that is, enhanced genetically to "have human intellectual capacities." That's a way, he says, of proving that "personhood not hum should any animal DNA ever be permitted to re-engineer human embryos?

Such experiments are far from unthinkable. A new social movement called "transhumanism" advocates the creation of a "post-human species," which would include using animal genes in progeny to increase strength or make senses more acute.

Once having children was generally conducted in a simple way: Men and women got married, made love, and had babies although not always in that order.

To the delight of some and the dismay of others, human reproduction has become a far more complicated matter.

Infertile couples now conceive through in vitro fertilization. Women who can't carry a child can arrange to have their baby gestated by a surrogate birth mother. Gay and lesbian partners are demanding the right to marry and have families. Sex selection has already begun.

The social controversies raised by these behavioural changes, already white-hot, are going to grow even more intense as cutting-edge procreative advances offer ever-greater latitude to those wanting children and more-precise control over the kind of children they have.

But critics worry that our growing mastery over reproduction could slide from liberty into license and even into reproductive anarchy. Look for these issues to cut through the body politic like a laser in the coming decades:

> *Will it be acceptable for a woman to use animal or artificial wombs to gestate her baby so as not to have her professional life inconvenienced by a wanted pregnancy?*

Is there a right to have genetically related offspring? Reproductive cloning is off the table for now because cloning isn't safe. But what if it were? Some bioethicists are already suggesting that outlawing reproductive cloning, at least for gay or infertile couples, would be unconstitutional because "procreative liberty" includes the right to have biologically related offspring.

Eradicating genetic disease is one thing. But there is a chorus of advocates who want to harness our growing knowledge of the human genome to "improve" our children through germ line genetic manipulations.

James Watson, co-discoverer of the DNA double helix, argued that prospective parents should use biotechnology to eradicate undesirable traits or fabricate wanted enhancements in their children.

More bluntly, bioethicist Gregory Pence suggested in "Who's Afraid of Human Cloning?" that parents should be allowed to use biotechnology to "aim for a certain type" of child "in the same way that great breeders try to match a breed of dog to the needs of a family."

Will the artificial womb affect the right to abortion?

Within 20 years, it is expected that artificial wombs will be available to save troubled pregnancies. If so, it is foreseeable that some anti-abortion states would pass laws requiring women wanting mid- or late-term abortions to instead have their unwanted fetuses saved through transfer into artificial wombs.

Science is opening the door to procreative "quality control." Whether society will permit made-to-order children will be the subject of debate for years to come.

All this begs the question: Who decides?

Some believe scientists should have the exclusive say because of their unique expertise. Thus, bioethicist Rahul Dhanda, wrote in "Guiding Icarus," that science "knows what is good for society, like a parent knows what is good for the child."

Professor Francis Fukuyama, a noted public intellectual, took a different view in "Our Posthuman Future."

"True freedom means the freedom of political communities to protect the values they hold most dear," he wrote, "and it is that freedom that we need to exercise with regard to the biotechnology revolution today."

The intellectual foundation is being laid for lawsuits that will seek a constitutional right to conduct scientific research perhaps in response to passage of a new federal law outlawing all human cloning.

Some scholars believe such a right is contained in the First Amendment. In this view, scientific experimentation is analoguous to a reporter's right to research a story.

If there is a right to conduct experiments implicit in the First Amendment, only a compelling state interest such as preventing a plague would justify the government infringing a scientist's fundamental freedom of inquiry.

Opponents of unfettered research say the scientist-equals-a-reporter analogy fails because granting a right to research would actually be akin to allowing a reporter to set fire to a building so he could write a story about the arson.

How all of this will turn out, nobody knows. But as Leon Kass, former chairman of the President's Council on Bioethics, has said: "All of the natural boundaries are up for grabs. All of the boundaries that have defined us as human beings, boundaries between a human being and an animal on one side and between a human being and a super human being or a god on the other. The boundaries of life, the boundaries of death. These are the questions of the 21st century, and nothing could be more important."

Biological Research Advances Agriculture

Now, the U.S. agriculture sector can reap more technological benefits resulting from biological research under the direction of the U.S. Department of Agriculture (USDA).

Agriculture research has produced disease testing for livestock, speciality-crop varieties and GPS technology for farm equipment to name a few.

The Agriculture and Food Research Initiative (AFRI), sponsored by USDA and created within the 2008 Farm Bill, established a competitive grant program to provide funding for fundamental and applied research, extension and education in food and agricultural sciences with a focus in six development areas. Grants that emphasize sustainable agriculture are preferred.

Priority Issues for Grant Applications:

1. Plant health and production and plant products
2. Animal health and production and animal products
3. Food safety, nutrition and health
4. Renewable energy, natural resources and environment

5. Agriculture systems and technology
6. Agriculture economics and rural communities.

"By focusing our resources on achievable and measurable outcomes, USDA's investment in science will help address some of America's and the world's intractable problems," said USDA chief scientist and director of USDA's Nation Institute of Food and Agriculture (NIFA) Robert Beachy.

"Today's announcement demonstrates USDA's commitment to supporting research, education and extension to bring about true change in areas like climate change, obesity and bioenergy."

In addition to the six priority issues for grant funding, the 2008 Farm Bill added conventional (classical) plant breeding, conventional (classical) animal breeding, renewable energy, domestic-marketing strategies and rural entrepreneurship as grant focal points.

There are also five designated primary challenge areas around which AFRI is structuring the grant program.

Challenge Areas:

1. Keep American agriculture competitive while ending world hunger
2. Improve nutrition and end child obesity
3. Improve food safety for all Americans
4. Secure America's energy future
5. Mitigate and adapt to climate change.

Agriculture Sec. Tom Vilsack allotted $262 million to this effort. State agricultural experiment stations, colleges and universities, university-research foundations, other research institutions and organisations, Federal agencies, national laboratories, private organisations or corporations and individuals are eligible to apply for grants. Grants are awarded on a 10 year basis and some grants are eligible for renewal upon conclusion of the term.

The benefits of research and technology to society cannot be quantified. The USDA has stepped to the plate to motivate the sustainability, efficiency, profitability and development of one of the nation's most economically providing sectors.

Do you agree that the selected issues and challenge areas should be addressed, or are there other issues? Should more or less money be awarded for research grants? Do you think the research will result in significant technological gains?

In September 2009, the AHB presented a proposal to Agriculture and Forestry Minister David Carter to amend the National Bovine Tuberculosis (TB) Pest Management Strategy. The strategy amendment was duly approved and the revised strategy came into effect on 1 July 2011. This strategy will guide the TB control programme through to 2026, subject to five yearly reviews.

Over the next 15 years, the strategy aims to achieve the following primary objectives.

These include:

> *The eradication of TB from wild animals over at least 2.5 million hectares of Vector Risk Area (VRA), including two extensive forest areas representing relatively difficult operational terrain from which to eradicate the disease*

Continued freedom from infection in wild animals (vectors) in existing Vector Free Areas (VFAs) and areas where eradication is considered to have been achieved.

A secondary objective is to maintain the national infected herd period prevalence level (the number of herds with TB during a period of time) below 0.4 per cent during the term of the strategy. The amended strategy gives priority to wildlife TB eradication and allows the AHB to priorities operations and resource allocation for this purpose.

The TB control programme has made significant gains over the past decade, especially in reducing the number of infected cattle and deer herds. However, TB-infected possums continue to be a source of livestock infection across some 10 million hectares of New Zealand's TB Vector Risk Area. The revised strategy sets out to address this underlying problem by aiming to eradicate TB from possum populations in selected areas. These areas make up 25 per cent, or 2.5 million hectares, of the total area of New Zealand known to contain infected wild animals. Achieving this objective will also confirm that TB can be eradicated from possums and other wild animals across large forest tracts where possum control is most challenging.

Eradicating TB from the possum population across one quarter of the total area known to be at risk from TB-infected wild animals would also from a basis for extending the eradication approach to further large areas of New Zealand.

The revised strategy will continue to protect the reputation and value of New Zealand's dairy, beef and deer exports by ensuring infected herd numbers remain below a 0.4 per cent period prevalence. To achieve the objectives of the revised strategy, the AHB will vigorously pursue improvements in the cost-effectiveness of possum control. Herd testing and movement control policies will also be adjusted to reduce the risk of herd-to-herd TB transmission and, over time, reduce the need for herd TB testing in areas of low disease risk.

With a TB incidence of below 0.4 percent, NZ is intending to eradicate the disease risk from their wildlife reservoir, from 2.5 million hectares. 25 per cent of the total area of NZ. With a TB incidence of almost 8 percent in the first half of 2011, Defra is 'mindful' of setting up a couple of pilot 150 sq km plots for a four year badger culling 'trial'. But using a published operating protocol which should guarantee the outcome of this plan is similar to that of its previous exercise in prevarication, the ~~RBCT~~ Badger Dispersal Trial.

A primary school is under threat of collapse after badgers tunnelled underneath it and shifted 7.5 tonnes of soil. The school, which teaches 420 pupils, has been blighted by the badgers - leaving deep gaps in the foundations underneath two main classrooms. Parents and governors now fear buildings could cave in at Ashcombe Primary School, Western-Super-Mare, Somerset.

A spokes~~man~~ person explained:

Our major concern is the fact that badgers have built or excavated under one of the buildings, which contains two classrooms. "These are elderly pre-fabricated buildings which need constant repair. The excavations by the badgers will have had some effect on the foundations.

"We worry that the buildings will collapse into the holes that have been left.

"Another problem is that the animals have brought their kill under the buildings, which of course we can't get to, so they decompose. All you can do is open the windows:

> *"Anodyne and jargon-spattered response from the department, which talks about 'new functionality' [* snip*] and seeks to paper over cracks rather than come clean about its shortcomings."*

The comment we note, is read from the same hymn sheet as that given to the FW, three weeks ago.

More of the problems with SAM are described in Farmers Guardian this week in a small snippet entitled 'IT Problems Hit Calf Exports.

The piece is as follows:

> *"Computer problems at the AHVLA Central Operations for Exports in Carlisle have seriously disrupted calf exports and it could be weeks before the backlog is cleared.*

Exports were suspended on September 28th and a statement sent to *Farmers Guardian* by the AHVLA said a new computer system was being developed to improve customer services, but problems in a new part of the system meant the agency was unable to establish the disease status of the some cattle prior to their planned export.

Although the article specifically mentions calf exports, this problem would affect all cattle exports, with breeding stock also snarled up. The window for exporting after a clear TB test is just 28 days from jab day, substantially less that the 60 day home market timeline, and export office vets at Carlisle would need to check the status of the consigning herd, as well as the log jammed test charts for these animals. And that apparently, they cannot do.

AHVLA problems with SAM are only one in a series of governmental cockups introducing new IT systems. The Telegraph describes a £12 billion bungle with NHS systems, following hard on the heels of the Fire Service's abandoned flagship which ratcheted up costs of £469 million of taxpayers' (borrowed) cash before being moth balled. The Telegraph piece on the NHS system notes that:

> *". last month, the MPs on the Public Accounts Committee described the system (cost: £12 billion to date) as "unworkable". But all is not lost. While angry farmers were quick to point out on a BBC TV report, their own particular problems, snarl ups and confusing paperwork, ROD (perleese don't say Rod Who?) - Defra's acronym 'ROD' is the Regional Operations Director - a very nice man called Mark Yates appeared on camera to say that if farmers are experiencing any problems with SAM, they should report them directly to him. And in particular, Defra's new toy, SAM. This system was due on stream in July, but three months on is apparently causing havoc. Farmers Weekly reported this snippet last week.*
>
> *Under the strapline 'Computer Problems Jeopardise TB processing', the FW report describes: how problems with the new [SAM] computer system and the loss of experienced administrative staff, have combined to create severe difficulties in dealing with TB paperwork."*

While inexperienced and agency staff struggle to input TB test results, the report highlights a log jamof over 700 TB test charts for the SW region (to the end of September) which have to be inputted manually."

The piece describes the morale of staff expected to 'lie to farmers' and the 'extra stress caused to lower grade staff'.

The report also describes problems with TB licensing experienced by one farmer, who having applied to move 20 animals, was given a licence for 23 including three breeding bulls, two of which had been sold. The final license sheet listed just 18 animals. So it would appear that manual overrides could be difficult, if not impossible too.

But with a unique piece of airbrushing, an AHVLA spokesperson gave the following comment to FW about Defra's new all singing, all dancing SAM computer system.

"This will improve the level of service provided to customers"

Not if the goddamn test charts aren't inputted on time, it won't. The main frame computer will generate a bloody shut down notice. Been there. Done that:

> *However minor issues have been identified, and have now been fixed"*
>
> *Really? How minor? And that backlog has decreased then? Yes? No?*

No. Our information is that it has increased. The purring continues:

> *"During this time, additional manual checks have been implemented on all TB test charts and associated processes to ensure AHVLA is able to identify and remove reactor animals from farms as quickly as possible."*

Additional manual checks? Why? So there is no backlog of reactors or test charts then?

According to many farmers we speak to, yes, there is. Mind, if the source was dealt with there wouldn't be any need for pernicious 60 day testing, abattoir slots, transport, or unskilled agency staff trying to log a million unique eartag numbers into a new computer system. But let that pass. Sadly, all this sounds a tad familiar. Remember the phrase 'don't mention the war' ? We see a distinct parallel, and having been on the receiving end of a syrupy recorded message (twice) recently when trying to contact AHVLA, have a sneaking suspicion they may be turning the phones off. For hours at a time. Nah, they wouldn't do that would they? What bit of 'Animal Health' is by passing the radar of Defra's top brass?

Of course they would. But don't forget, the AHVLA lady said it will " improve the level of service." What she did not say was when is a well known cause of blood pressure when patients visit a doctor's surgery; the sight of pristine, dazzling white attire, guaranteed to cause a flutter in most of us.

So imagine the reaction of cattle, more used to olive green - be that grass, silage or the boss's overalls - faced with the eye catching vision of a man in a white coat advancing.

Like it, they do not. As was found a decade ago during the carnage of FMD.

The picture below, is by Chris Chapman and was taken on a Dartmoor dairy farm in 2001. Even normally placid dairy cows are looking askance at the ghostly figure as they are rounded up to be shot. These cows were sedated. Cattle gathered for inspection and TB testing by AHVLA operatives are not.

Or that was how it used to be until the arrival of Defra's new, all singing, all dancing and very streamlined central ordering system. This was designed to give economy of scale, with orders channelled through the Finance section of the regional office. We understand that since FMD, Defra staff are not allowed to use the infamous 'white overalls' in the field. This is partly to avoid media surveillance but mainly to avoid being squashed by freaked out cattle.

But guess what? Our moles tell us that what were delivered by a this new team of desk bound admin clerks (with no idea of the problems faced at the coal face and no remit to listen) were packs of tissue-paper thin, single use overalls. And they were white.

"Under this new program, if a wild free-ranging, white-tailed deer harvested during the 2011 hunt is turned in for bovine TB testing, and it cultures positive, the hunter may apply for a $200 incentive," said MDARD Director Keith Creagh.

Bovine TB is a contagious bacterial disease of cattle that can affect other mammals, including humans. In 1994, a unique strain of bovine TB was identified in Michigan's free-ranging deer.

"MDARD recently announced 57 counties in Michigan's Lower Peninsula achieved bovine TB-free status; but there is still a pocket of bovine TB in deer that can be transmitted to cattle," Creagh said.

"This new incentive program is one tool in our toolbox to help refine the footprint of the disease and protect Michigan's $9.2 billion beef and dairy industries."

"Some of the best hunting in the state is in Northeastern Lower Michigan," said State Sen. John Moolenaar (R-Midland). "Our wildlife enthusiasts can show they care about TB eradication, and at the same time, Michigan's Department of Agriculture and Rural Development will reward them for removing disease from the landscape."

Under the Incentives Program:

Hunters must take the deer to a DNR check station.

They can have the antlers removed, but the head is submitted for testing.

DNR collects the heads from all the check stations and transports them to the Wildlife Disease Laboratory at Michigan State University's (MSU) Diagnostic Centre for Population and Animal Health.

If the deer is confirmed to be TB infected, the hunter will receive routine test notification from the DNR laboratory which will include an Incentives Program contact number.

Notified hunters should contact MDARD with their confirmation code. A form will be mailed to the hunter's address for them to fill out and mail back for payment.

Upon receipt of the completed form, hunters will be mailed $200 for each TB-positive deer harvested.

Visit the Emerging Diseases web site for an all-positives map and additional information, including an advertisement about the incentives program; or to join the Michigan Animal Health LISTSERV.

In the time period from one year after the last proactive cull to 28 August 2011 (the post-trial period), the incidence of confirmed breakdowns in the proactive culling trial areas was 28.0% lower (95% CI: 15.0% to 39.1% lower) than in survey-only areas, and on lands up to 2km outside proactive trial areas was 4.1% lower (95% CI: 25.7% lower to 23.7% higher) than outside survey-only areas.

And continues;

Exploratory analyses stratified by 6 month periods are consistent with an ongoing, but diminishing (test for temporal trend $p=0.008$), benefit of proactive culling continuing through the latest 6 month period analysed (55 to 60 months post-trial).

The paper also observes that "the effects observed outside trial areas remained consistent with no ongoing effects of proactive culling in these areas."

Stating that their "post-trial results must, of course, be considered in the context of the smaller reduction seen inside proactive trial areas and the increased incidence seen outside proactive trial areas in the period from the end of the initial proactive cull until one year after the last proactive cull in each triplet."

Research Platforms at the Donnelly Centre

The human body is a finely tuned machine, powered by genes, proteins and cells that operate in a complex, efficient system. Traditional biology studied these components one at a time, yielding limited insights about the way a machine functions. It is now apparent that the most effective way to understand how the human body works, and to best repair it when it breaks down, is to unravel how these different components work together.

The Donnelly Centre takes this holistic approach to biomedical research: By integrating technology, expertise and thought from a diverse array of disciplines, it aims to unravel some of the great complexities of biology. In the process, the Centre is poised to make some significant future advances in medicine and health.

At the Donnelly Centre, biomedical research is defined by three broad platforms:

Integrative Biology

Includes: Systems Biology; Functional genomics; Proteomics; Bioinformatics; Chemical Biology and Chemical Genomics; Structural biology.

The Donnelly Centre studies the foundations of biology using an integrative approach. Researchers with diverse perspectives and scientific backgrounds work together to investigate the complex ways that genes, proteins and small molecules interact. Their studies span the following disciplines:

Systems Biology: studies a living organism by viewing it as an integrated and interacting network of genes, proteins and biochemical reactions. Instead of analysing an individual cell nucleus, for example, systems biologists focus on all of the components of the cell and the interactions among them, as part of one system.

Functional Genomics: uses the vast amounts of genomic data to describe gene (and corresponding protein) functions and interactions.

Proteomics: is the large-scale study of the structures and functions of proteins, the main components of the physiological metabolic pathways of cells. The term "proteomics" was coined to make an analogy with genomics, the study of genes.

Ribonomics: is the large scale study of RNA functions, metabolism, movements and structures. New research is suggesting that RNAs play an equally important role as proteins in the control of cellular architecture and function.

Bioinformatics: uses information technology to study biological processes. Recent advances in biological research have led to an explosive growth in scientific data. Bioinformatics develops and advances computerised databases, algorithms and statistical techniques to best store, manage and analyse all this biological information.

Chemical Biology and Chemical Genomics: probe living systems at the chemical level. Chemical biology uses chemical compounds, or small molecules, to study and manipulate biological systems. Chemical genomics studies how genes respond to small molecules.

Structural Biology: elucidates the molecular structure of biological macromolecules, such as proteins and nucleic acids, and studies how they acquire their structures, and how their function is influenced by their structure.

Bioengineering and Functional Imaging

Includes: Regenerative medicine; technological developments in cell and tissue imaging; high-throughput cell biology.

No holistic approach to biomedical research would be complete without a cell biology platform. Several researchers at the Donnelly Centre study the dynamics of healthy cells, while others investigate abnormal ones, such as cancer cells. A greater understanding of the cellular basis of disease helps to develop improved methods of diagnosing disease and more effective treatments for disease. This research draws upon, and informs, the following disciplines:

Regenerative Medicine: harnesses the power of stem cells to repair, regenerate or replace diseased cells, tissues and organs.

Technological Developments in Cell and Tissue Imaging: Recent advancements in imaging technology stem from nanotechnology (e.g., metal-based nanostructures), microtechnology (e.g., micro-sized machines), and molecular engineering (e.g., bacteria-infecting viruses that connect proteins with the genetic information that encodes them).

High-throughput Cell Biology: uses a combination of automation equipment and classical cell biology techniques to perform rapid, large scale research. It addresses questions such as: how cells function and interact with each other, and how pathogens exploit cells in disease.

Models of Disease Includes: Stem Cell Biology; Model Organisms; Animal Models of Human Disease

The Donnelly Centre also uses model organisms, such as yeast, and stem cells to gain a better understanding of how diseases arise and develop. This research helps to identify new methods for prevention and treatment. It encompasses:

Stem Cell Biology: studies how human embryonic stem cells develop into malfunctioning cell types (for example, insulin-producing pancreatic cells or heart muscle cells).

Model Organisms: include yeast – an organism that shares roughly one third of genes with humans – as well as worms and mice. Studying the genetic interactions that cause disease in simple organisms, for example, sheds light on the genetic basis for disease in humans.

Animal Models of Disease: is the study of a non-human animal that has a disease or injury similar to the human condition under investigation.

Biological Impact Assessment of Direct CO2 Injection into the Ocean

Practical implementation of CO2 sequestration through direct injection of anthropogenic CO2 into the deep ocean could have a significant environmental impact on marine organisms. In this paper, we summarise recent advances in biological research for direct CO2 injection and discuss required research work for the future. Possible biological impacts caused by CO2 storage in the ocean can be categorised into acute and chronic. Acute impacts such as on the survival of marine organisms could be determined by laboratory and field experiments and assessed by simulation models. Chronic impacts, on the other hand, such as sublethal effects would be difficult to verify by the same approach as for acute impacts. One of the practical solutions for this issue is field experiments starting with controlled small-scale and, eventually, large-scale CO2 injection intended to determine ecosystem alteration. Practical implementation of CO2 ocean sequestration must employ a comprehensive monitoring program of its biological impacts. The ultimate criteria for the actual implementation of marine CO2 injection will be based on comparison of the biological responses of ecosystems affected by CO2 between the euphotic zone and the deep sea. Assessment is urgently needed of the comparative differences in marine-ecosystem alterations due to impacts to oceanic euphotic zones induced by the predicted increase in ambient CO2 levels in the atmosphere over the coming decades and impacts to the deep ocean through direct CO2 injection to sequester anthropogenic CO2.

Intoduction

Continuous discharge of CO2 into the atmosphere will not only accelerate global warming but also raise the partial pressure of CO2 and lower the pH in the ocean, both of which consequently may result in biological impacts [1, 2, 3]. On the other hand, feasibility studies recently suggest that CO2 sequestration by direct injection of anthropogenic CO2 into the deep ocean could help reduce the atmospheric CO2 concentration and mitigate global warming [4, 5]. Based on the injection depth of CO2, there are two different principal types of the sequestration technology, i.e. dispersion of CO2 at intermediate depths (mid-depth type) and injection of CO2 at depths greater than 3,000 m (lake type) [6]. Irrespective of differences in storage schemes, their implementation could have a significant impact on deep-sea organisms [7]. The environmental/biological impact is the most essential and controversial problem of direct CO2 ocean-storage technologies. Before implementation of such CO2-emissions mitigation efforts, a thorough environmental impact assessment is necessary. In this paper, we discuss practical ways of implementing biological impact assessments of direct injection of CO2 into the deep ocean and indicate the required research strategies.

Acute and Chronic Impacts

Possible biological impacts caused by marine CO2 addition can be categorised into acute and chronic. The duration of an acute aquatic toxicity test is generally four days or less and mortality is the response measured. In contrast, a chronic aquatic toxicity test often requires periods from several weeks to years in length and a biological response such as growth is the response measured [8]. When considering direct injection of CO_2, it can be said that acute impacts will mostly occur at and near the CO_2-injection point, whereas chronic impacts will extend over a large area.

Acute Impacts

Acute impacts such as the effect of elevated CO_2 concentration on the survival of marine organisms can be determined by laboratory [9, 10] and field [11, 12] experiments and assessed by simulation models [13 to 16]. This approach has been developed through the Japanese R&D program [3] although its practicable assessment needs further studies such as those outlined below.

Careful Consideration on Sensitivity-to-CO_2 Differences between Shallow and Deep-Sea Species Most CO_2-toxicity tests carried out so far have examined the effects on epipelagic species. Development of experimental procedures for deep-sea (mesopelagic, bathypelagic and abyssopelagic) species is needed. There are two basic approaches to conduct such experiments: laboratory and *in situ* studies. A few laboratory [17] and *in situ* studies [11, 12] have been reported. Laboratory studies of living deep-sea organisms require their capture, recovery, and maintenance, and experimental apparatus replicating the deep-sea environment. Laboratory studies have an advantage in that they allow precise control of CO_2 concentration during experimentation (*in situ* studies do not allow this control). Sensitivities of deep-sea species to high CO_2 may be predicted by extrapolation from existing data of well studied epipelagic species, based on the mechanism of the CO_2 effect.

Comparison of Tolerance to High CO_2 Levels and Its Mechanism among Species and Their Developmental Stages Tolerance of marine organisms to high CO_2 concentrations significantly differs among species and among their developmental stages [9 to 12, 18]. Susceptibilities of marine organisms to high CO_2 are highly relevant to their compensation mechanism. It appears that most invertebrates lack the ability of acid-base regulation to compensate for acidosis associated with high CO_2, whereas fish (vertebrates) have well established compensation mechanisms.

Since we have only limited data of CO_2 tolerance of marine organisms, experiments should be conducted on a wide range of taxa. The tolerance should also be examined for various developmental stages of organisms, since their physiology drastically changes with development. A very small decrease in survival during early developmental stages has been shown to cause a large decrease in recruitment success, i.e. significant alteration in population.

Although organisms encountering the high CO2 plume created by CO2 injection will experience various CO2 concentrations for variable durations, the majority of CO2-toxicity tests so far have been carried out with steady, constant CO2 concentrations (as is generally the case with conventional toxicity tests). Our recent experiments indicated that mortalities are significantly different between fish exposed to steady and variable CO2 concentrations. When fish were immediately exposed to CO2 of 7 kPa water from normocapnic water (steady exposure), all the fish died instantaneously (100% mortality, upper graph). However, mortality of fish was only 30% under a stepwise increase of CO2 to 7 kPa.

In addition to that, the rapid return to normocapnia caused a drastic increase in mortality. We hypothesize that the instantaneous death of fish under both a rapid increase and decrease of CO2 concentration was caused by a disruption of the nervous system associated with acidosis and alkalosis respectively. The lowered mortality under the stepwise increase of CO2 can be explained by the compensation mechanisms of acidosis through acid-base regulation [9]. Responses by the nervous system occur on a time scale of seconds, whereas acid-base regulation via various physiological mechanisms requires a time scale of at least minutes. Toxicity tests with variable CO2 concentrations are needed for precise evaluation and prediction of acute impact consequences. The effects of variable ambient CO2 concentrations should be experimentally re-examinedthus avoiding the uncertainties of extrapolating from constant-CO2 experiments [16, 20]. Differences in mortality between fish (*Sillago japonica*, demersal fish of coastal/shallow sea, mean total length ?S.D. = 36.8 ?6.2 mm) exposed to steady and variable CO2 concentrations (N = 20 for each experiment, Kikkawa, Kita and Ishimatsu, unpubl. data). Upper graph: Mortalities of fish under immediate exposure to CO2 of 3 kPa (Ï%), 5 kPa (²%) and 7 kPa (%) water from normocapnic water (steady exposure).

Lower graph: Fish were exposed from normocapnic water to CO2 of 1, 3 and 5 kPa water for 2 hrs in sequences, and to 7 kPa water for 12 hrs, finally to normocapnic water immediately (solid line, variable exposure). Circles (Ï%) with a line indicate mortalities corresponding to variable exposure. Note that the mortality rapidly increases with the immediate return to normocapnia.

Chronic Impacts

Chronic impacts of direct injection of CO2, i.e. sublethal effects of CO2 on the population dynamics and biodiversity that are directly related to the marine ecosystem, would be difficult to verify by means of laboratory experiments and to assess using ecosystem models. One of the proposed solutions for this issue utilises field experiments starting with controlled small-scale, and eventually large-scale, CO2 injection intended to determine ecosystem alteration. Such field experiments can be useful for confirming expected consequences from laboratory experiments and avoiding the uncertainties of extrapolation.

Small-scale field experiments may not be suitable for biological impact studies of mid-depth CO2-dispersion technology. Since 1) the biomass of the mid-ocean is significantly lower than that of the ocean-floor, and 2) the released CO2 can rapidly dilute to about 1/50,000 its initial concentration [21], the ecosystem impact would not be at a detectable level within a small-scale experiment. The scale of a field experiment required for detecting ecosystem alteration induced by mid-depth CO2 dispersion would be on nearly the same scale as the actual implementation scale of the technology. As regards lake-type CO2 injection, sluggish and immobile organisms inhabiting the area covered by a liquid CO2 (lake) would have no chance of survival. The CO2 plume induced by dissolution of the CO2 lake would impact the ecosystem in the area around the lake. The ecosystem impact caused by lake-type technology can be assessed by means of small-scale field experiments [11, 12, 22].

As far as species extinction at local and within-short-time scales is concerned, mid-depth CO2 dispersion would cause less harm than lake-type CO2 injection. However, lake-type injection of CO2 into restricted depressions of the seafloor has an advantage in that the biological impacts will be limited to within a relatively local area [23]. Because the scale of the consequences and quality of biological impacts differ between these methods, precise evaluation of which strategy would cause less overall harm needs information about the specific quantity of CO2 to be injected, which will be determined from economic and technological analysis of these mitigation options, Determination of Ecosystem Alterations

Although a broad geographical range of deep-sea ecological studies are required for discussion of the feasibility of ocean CO2 storage, it may be most efficient and effective to start a detailed research project at a regional scale once specific CO2 injection sites have been proposed.

Such a project, whether small or large scale, would need to start by accumulating baseline information through a deep-sea field survey of biomass, biodiversity [24] and the trophic structure together with collection of physicochemical data, in order to evaluate induced ecosystem alterations. These alterations would be reflected in the food-web structure through changes in the trophic position of organisms. For example, the trophic position of certain omnivorous feeders that feed equally on both lower and higher trophic organisms would shift if their food composition rate of lower/higher trophic organisms were altered by CO2 injection.

An example of trophic relationships of mesopelagic organisms (Kita and Mito, unpubl. data). Sample organisms were collected from 500 m to 1500 m depth in the sub-arctic area off Japan (36.41N, 141.90E). Upper graph: To avoid variance in ´13C due to lipid content levels of the organisms, the solvent-extractable lipid fraction was removed from samples for ´13C analysis. Symbols indicate copepods (Ï%), decapods (²%), salps (*) and fish (_). Lower graph: Estimation of the trophic level (TL) of copepods, decapods and fish using their ´15N values. Enrichment value of +2.7 ‰ for ´15N was used to predict trophic positions. It should be noted that knowledge of deep-sea ecosystems is fragmentary due to the logistical difficulties of deepsea research. Therefore analysis of such ecosystems must take notice of new scientific discoveries e.g. the possible high abundances and biomass of gelatinous zooplankton (siphonophores, medusae, ctenophores and salps) in the mesopelagic zone [25] and the importance of the marine microbial ecosystem (microbial-loop) in the physical cycle The current status of deepsea ecosystem modelling is still immature because sufficient observational data sets are lacking.

Although modelling ecosystem consequences that result from CO2 additions would be difficult, modelling tools are useful for gaining an understanding of ecosystem structures and mechanisms. Developing a deepsea, especially a mesopelagic, ecosystem model is also required for the further understanding of the carbon cycle in the ocean. It must be kept in mind that the disappearance of certain species may not influence the overall carbon flux but would nevertheless represent a huge loss to biodiversity.

An example of mesopelagic ecosystem model based on the carbon flux. The microbial loop is the main focus in this model. Micrograzers and mesograzers include ciliates and copepods respectively. Recently an assessment method for the extinction risk to populations induced by toxic chemical compounds has been developed [27 to 29]. Such an approach to ecosystem assessment may be suitable for evaluating the impacts of direct CO2 injections.

Monitoring of Direct CO2 Injection

As discussed above, acute and chronic impacts on marine ecosystems can be assessed by means of laboratory- and field-experiments and numerical models. However, we cannot avoid pointing out that there is a possibility that environmental impacts of marine CO2 injection will be larger than we expect. Therefore practical implementation of CO2 ocean sequestration must, as a precautionary measure, employ monitoring programs for its biological impacts. The monitoring programs should be designed for both acute and chronic impacts. Those for the acute impacts will be restricted to areas near CO2-injection points and need to be conducted at frequent intervals, whereas those for chronic impacts will be extended over much larger areas, up to basin scale. The area and the frequency of monitoring programs can be configured by numerical models based on the diffusion of injected CO2.

Concluding Remarks

The ultimate criteria for the actual implementation of CO2 injection into the ocean will be established through comparison of responses of the CO2-affected ecosystems from the ocean surface (impacted by increased atmospheric CO2 concentration) to the deep sea (impacted by CO2 injection). Such a comparison based on biological data is urgently needed. It should be taken into account that a considerable amount of the protein source for the world population depends on the marine epipelagic production. It is now well accepted that surface ocean pCO2 levels will double over their pre-industrial values by the middle of this century and that the marine ecosystem will therefore be impacted [30] regardless of whether or not CO2 ocean sequestration is implemented.

Advances in Biological Psychiatric Research

The 19th century writer Lewis Carroll wrote “Alice in Wonderland”. In this children’s classic was a delightful character called ”The Mad Hatter”. This strange character wore a 19th century hat. The character was based on an actual situation in the felt hat industry in England. It seems that the people who made felt hats were using mercury in the process and were getting mercury poisoing.

Mercury is a heavy metal that poisons the brain causing madness. But what does this tell us? For one thing, we need to be very careful about pollution. For another thing, a toxin is capable of poisoning the brain and therefore schizophrenia might be caused by an unknown toxin. Alcohol is an example of an exogenous toxin that poisons the brain. However, in schizophrenia an endogenous toxin may be at work. There even could be more than one toxin.

Catecholamines

The catecholamine theory is one of the most popular theories for schizophrenia and also one of the most logical. It takes into account the connection between schizophrenia and stress. Stress exacerbates schizophrenia as well as other psychiatric diseases. Ref. 1 describes one of the many catecholamine theories.

Dopamine is one of the suspects (2). It seems that neuroleptic drugs block dopamine receptors, which is why they produce Parkinsonian side effects. However, dopamione itself may not be the problem in schizophrenia. A toxic metabolite or metabolites may be the problem(s). Hoffer favours this theory, as did the late Dr. Osmond.

Axelrod was awarded a Nobel Prize for his work on this theory (4). Unfortunately he was never able to identify the toxin.

COMT

This very important enzyme metabolises dopamine by methylation. Supposedly this deactivates dopamine, but there is a theory that abnormal methylation can produce a toxin (5, 6). A COMT inhibitor has been used as a treatment in Parkinson's disease (9). In theory it makes the dopamine last longer.

Drug Abuse

A lot can be learned about the brain when studying drug abuse (10). Many hallucinogens have chemical structures similar to metabolites of dopamine. Dopamine itself has no methyl or methoxy groups. The hallucinogens have either methyl groups or methoxy groups or both. Some hallucinogens have structures similar to serotonin. This may be a reason why so many SSRIs (selective serotonin reuptake inhibitors) are being used in psychiatry.

The class of hallucinogens similar to dopamine in structure is called "phenethylamines" (11). Amphetamines are classified by the chemical name phenylisopropylamines (12). An "amine" by definition contains the nitrogen atom.

Unfortunately some of these hallucinogens can lead to physical dependence (addiction).

The serotonin theory goes back to Gaddum (16) and others. According to this theory there is an antagonism between serotonin and LSD. This led to the serotonin deficiency theory for schizophrenia. However, actual measurements of serotonin postmortem in schizophrenic brains did not support this theory. A similar theory was advocated for depression.

Drug abuse is such a vast topic that I can't do justice to it here. Perhaps I will write an article strictly about it, but even that won't do justice to it. I would have to write a whole series of articles on it like I have on schizophrenia.

Conclusions

My own theories are presented in Refs. 17-20. I believe that a toxin is created from dopamine by abnormal methylation. This abnormal metabolite could produce yet another abnormal metabolite. But what is the treatment? I am not at all satisfied with the drugs that we have now. I favour alternative medicine. I don't accept all types of alternative medicine, but orthomolecular medicine uses nutrition. That is real science.

Nutrition is safer than drugs and cheaper. Then again, I would not rule out drugs of the future. Maybe in the future they can do better than they have in the past.

During the past decade, rapid developments in single molecule techniques have allowed scientists to monitor and manipulate the position, orientation and distribution of individual molecules, as well as their dynamical processes, particularly chemical reactions, with nanometre spatial precision and millisecond time resolution. These developments have facilitated the advancement of many fields of biological research, and scientists can now observe and manipulate the behaviours of single biomolecules not only in vitro but also in live cells, gaining molecular level understanding of life processes.

BIOPIC aims at further developments and applications of single molecule techniques in order to study the dynamic processes of biological systems, including non-equilibrium dynamics of biomolecules as well as biochemical reactions in live cells, particularly, gene expression and regulation. New insights derived from these studies will advance our understanding of fundamental questions in biology and medicine.

High-throughput DNA sequencing has revolutionised biological and medical research, offering unprecedented opportunities to systematically dissect functional elements in genome sequences and gene regulation networks. BIOPIC will apply state-of-the-art high-throughput techniques to carry out functional genomics analysis of stem cells, human embryonic development and human diseases, especially cancers and infectious diseases. BIOPIC is capitalising on new sequencing techniques, with a focus on the single cell genomics, transciptomics and epigenomics analysis. These developments will greatly facilitate advances in biological research, medical diagnosis, and personal medicine.

Limited by light diffraction, conventional optical microscopy can only achieve spatial resolution of a few hundred nanometres. Recently, a few emerging super-resolution techniques have broken this barrier and achieved spatial resolution as high as 10 nm, and thus opened a broad range of applications in life sciences. BIOPIC uses these techniques to probe static structures, such as three-dimensional structures of nuclei; as well as dynamic processes in the living cell. The study will lead to unprecedented visualisation and deeper understanding of various life processes

Cells are the basic functional unit of life. Traditional studies of cell biology, which are based on the average results of a large number of cells, mask widespread cellular heterogeneity and therefore offer limited information. Microfluidics is a technology that manipulates liquid with the volume of nanolitre or less. Microfluidic devices suitably match a cell's dimension and hence provide a powerful tool to deliver and manipulate single cells under controlled microenvironments. At BIOPIC, we interface microfluidic technologies with high-throughput studies of real-time microscopy, as well as single-cell genomics and digital transciptomics.

Label-free imaging technology has become an important goal for microscopy. It generates contrast through a molecule's intrinsic properties, avoiding the disadvantages of fluorescent labels. BIOPIC is developing stimulated Raman scattering (SRS) microscopy for real-time and non-invasive biological imaging, with high specificity, high sensitivity and high spatial resolution. This technique is superior in observing the spatial distributions and real-time dynamics of many molecules, such as lipids, proteins, DNA, RNA, water, metabolites and small drug molecules, in live cells as well as in live animal and human tissues. SRS microscopy provides a unique tool for fundamental research in life sciences and medical diagnostics

Chapter 2

Biological Research in Marine Life

The aim of this special issue of Biological Research is to present some of the latest advances in biological research of scientists under the Frontiers of Science Program of the Chilean Academy of Sciences of the Instituto de Chile.

This program started 7 years ago, with the idea of incorporating into the Academy the very best young scientists working in Chile in order to discuss exciting advances and opportunities in their fields in a format that encourages informal collective, as well as one-on-one, discussions among participants. Several symposia have been organised over the years, and in each case speakers have been urged to focus their talks on describing current cutting-edge research in their disciplines to colleagues outside their field and to address questions such as: "What are the major research problems and distinctive tools of your field?" and "How might insights derived from other fields contribute to overcoming these limitations?" Participants remain active in the Frontiers of Science Program for an average of 3 years. Such groups have included leading researchers in disciplines such as astronomy, biology, chemistry, geology, mathematical sciences and physics. Here the Society of Biology has invited biologists that have been part of Frontiers of Science Groups to present their latest and most exciting work to the readers of Biological Research.

In the first study, Allende's team describes how a teleost, the zebrafish, can be useful to study the cellular and molecular mechanisms associated with copper toxicity in water, establishing a framework for further analyses of copper excess using an entire organism. The use of invertebrate models for research is also presented by Glavic et al. In this work, they describe how the stability of the Notch protein is dependent on the activity of GDP mannose dehydratase (GMD), and therefore on the levels of O-fucose moieties added to the Notch extracellular domain. These post-translational modifications will ultimately contribute to regulate the endocytosis and degradation of Notch.

The usefulness of vertebrate models for research is discussed by several papers describing immunology, cell biology, neurobiology and development. Thus, during the development of the nervous system morphogens are responsible for the establishment of neuronal territories. Palma, Larrain et al describe how the extracellular matrix may modulate the function of the morphogen Sonic Hedgehog (SHH). They specifically propose perlecan as the proteoglycan involved in both SHH localisation and activity, through a direct interaction among these molecules. After the establishment of neural territories, several cellular events lead to neuronal differentiation. Borquez and Gonzalez-Billault analyse the molecular mechanisms contributing to the establishing of cell polarity.

This review presents evidence relating the ubiquitin proteosome and calpain systems, with cell migration and the development of an axon in neuronal cells, two processes that require the polarisation of cells. Another interesting aspect of neuronal function is described in the work of Couve's team. These authors review the state-of-the-art in studying the endoplasmic reticulum in neuronal cells, providing a detailed analysis of its structure, composition, dendritic distribution and dynamics. A functional link is provided with the analysis of the possible roles for the endoplasmic reticulum in synaptic transmission. In the immunological system, dendritic cells (DC) are responsible for maintaining adaptive immunity because they can modulate T-cell activities.

Kalergis' team present a review of the interactions between dendritic and T-cells. Consequences of altered DC-T cells are analysed in the context of autoimmune diseases. Pathological conditions are also presented in the paper by Marzolo and Farfan. This article deals with the role of megalin, a receptor for apolipoprotein E that is implicated in several diseases such as Lowe Syndrome, Dent Disease, Alzheimer's disease and gallstone disease. The authors review the current known functions and propose mechanisms that control megalin expression.

Vegetal biology is also represented by the work of Moreno and Orellana, which describes the plant unfolded protein response (UPR). This is an up-to-date review showing the emergence of plant homologues for UPR proteins, and proposes some mechanistic linkage to physiological responses in plants.

Nature biologists are also represented in this special issue by three articles. Nespolo's team, analyse how morphological changes occurring during ontogeny may alter the metabolism of invertebrates. Metabolism in vertebrates is analysed by Ramirez-Otarola and Sabat. They present an article showing how dietary habits may influence the physiology of digestion in eight bird species. Changes in the enzymatic activities amongst species led the authors to propose a genetic base for such differences. As well, by analysing the distribution of breeding bird species, Abades and Marquet, characterise ecological phenomena. Based on statistical analyses of bird distribution, they propose that differences may be related to macroecological behaviour.

Finally, in a novel discipline, Gonzalez-Nilo et al. present a review of a new sub-discipline termed nanobioinformatics. The article shows some examples of nanobiotechnology in life sciences, and discusses new challenges for bioinformatics and computational chemistry.

We hope this number of the Biological Research *will be of interest for a broad number of biologists. We appreciate the energy and enthusiasm of the authors and other scientist that helped during the review process.*

Biological Research: Recent Advances in Angiotensin II

Angiotensin II (Ang II)* is a multifunctional hormone that influences the function of cardiovascular cells through a complex series of intracellular signalling events initiated by the interaction of Ang II with AT_1 and AT_2 receptors. AT_1 receptor activation leads to cell growth, vascular contraction, inflammatory responses and salt and water retention, whereas AT_2 receptors induce apoptosis, vasodilation and natriuresis. These effects are mediated via complex, interacting signalling pathways involving stimulation of PLC and Ca^{2+} mobilisation; activation of PLD, PLA_2, PKC, MAP kinases and NAD(P)H oxidase, and stimulation of gene transcription. In addition, Ang II activates many intracellular tyrosine kinases that play a role in growth signalling and inflammation, such as Src, Pyk2, p130Cas, FAK and JAK/STAT.

These events may be direct or indirect via transactivation of tyrosine kinase receptors, including PDGFR, EGFR and IGFR. Ang II induces a multitude of actions in various tissues, and the signalling events following occupancy and activation of Ang receptors are tightly controlled and extremely complex. Alterations of these highly regulated signalling pathways may be pivotal in structural and functional abnormalities that underlie pathological processes in cardiovascular diseases such as cardiac hypertrophy, hypertension and atherosclerosis.

Angiotensin II (Ang II) regulates blood pressure, plasma volume, sympathetic nervous activity and thirst responses. It also plays an important pathophysiological role in cardiovascular disease, including cardiac hypertrophy, myocardial infarction, hypertension and atherosclerosis. Ang II is produced systemically via the classical renin-angiotensin system and locally via the tissue renin-angiotensin system. Ang II was initially described as being primarily a vasoconstrictor peptide. However, recent studies demonstrate that Ang II has growth factor and cytokine-like properties as well.

At the cellular level Ang II modulates contraction, it regulates cell growth, apoptosis and differentiation, it influences cell migration and extracellular matrix deposition, it is pro-inflammatory, it stimulates production of other growth factors (e.g., platelet-derived growth factor, PDGF) and vasoconstrictors (e.g., ET-1), and it transactivates growth factor receptors (e.g., PDGFR, epidermal growth factor receptor (EGFR), and insulin-like growth factor receptor (IGFR)). The multiple actions of Ang II are mediated via specific, highly complex intracellular signalling pathways that are stimulated following initial binding of the peptide to its specific receptors. In mammalian cells Ang II binds to two distinct high-affinity plasma membrane receptors, AT_1 and AT_2.

The term "intracellular signalling pathway" includes the complex interrelated molecular cascades that transmit information from the membrane receptor to the intracellular proteins that regulate cell activities. The present review focuses on recent advances in Ang II signal transduction in cardiovascular cells. The growth factor and cytokine-like properties of this vasoconstrictor peptide will be highlighted and implications in cardiovascular disease will be discussed.

Angiotensin Receptors

AT_1 and AT_2 receptors, which are both seven transmembrane spanning G protein-coupled receptors, have been cloned and pharmacologically characterised. Pharmacologically the receptors can be distinguished according to inhibition by specific antagonists. AT_1 receptors are selectively antagonised by biphenylimidazoles such as losartan, whereas tetrahydroimidazopyridines such as PD123319 specifically inhibit AT_2 receptors (3,4). Two other Ang receptors have been described, AT_3 and AT_4 (5). However, the pharmacology of these receptors has not been fully characterised, and therefore these receptors are not yet included in a definitive classification of mammalian Ang receptors (6).

AT_1 Receptor

In humans there is only a single AT_1 receptor type, whereas in rodents, two subtypes of the AT_1 receptor have been identified (AT_{1a} and AT_{1b}). To date, AT_1 receptors have been shown to mediate most of the physiological actions of Ang II and this subtype is predominant in the control of Ang II-induced vascular functions (1). In the vasculature, AT_1 receptors are expressed mainly in smooth muscle cells (7). In the heart, AT_1 receptors are present in cardiomyocytes and fibroblasts (7). Ligand-receptor binding leads to activation of G proteins through exchange of GTP for GDP, resulting in the release of a and ßg complexes, which mediate downstream actions. AT_1 receptors interact with various heterotrimeric G proteins including Gq/11, Gi, Ga12 and Ga13. The different G protein isoforms couple to distinct signalling cascades. For example, Gq activation results in activation of phospholipase C (PLC), whereas GaI leads to cGMP formation. Although G protein-coupled receptors do not contain intrinsic kinase activity, they are phosphorylated on serine and threonine residues by members of the G protein receptor kinase family. AT_1 receptors are phosphorylated in the basal state and in response to Ang II stimulation. Various tyrosine kinases, including Janus kinases (JAK and TYK), Src family kinases, and focal adhesion kinase (FAK) can tyrosine phosphorylate AT_1 receptors (8).

AT_2 Receptor

The second major isoform of the Ang receptor, AT_2, is normally expressed at high levels in fetal tissues, and decreases rapidly after birth (9). In adults, AT_2 receptor expression is detectable in the pancreas, heart, kidney, adrenals, myometrium, ovary, brain and vasculature (9). In blood vessels, AT_2 receptors predominate in the adventitia and are detectable in the media. The AT_2 receptor is re-expressed in adults after vascular and cardiac injury and during wound healing and renal obstruction, suggesting a role for this receptor type in tissue remodelling, growth and/or development. The functional roles of AT_2 receptors are unclear, but these receptors may antagonize, under physiological conditions, AT_1-mediated effects by inhibiting cell growth, and by inducing apoptosis and vasodilation (10,11). Recently, in vivo studies of forearm blood flow in healthy human subjects found that the AT_2 receptor was not involved in modulating vascular tone in these subjects (12). Other studies suggest that AT_2 receptors also contribute to pathological processes associated with cardiac hypertrophy and inflammation (13). Signalling pathways through which AT_2 receptors mediate cardiovascular actions have recently been elucidated. Four major cascades are involved including 1) activation of protein phosphatases and protein dephosphorylation, 2) regulation of the nitric oxide-cGMP system, 3) stimulation of PLA_2 and release of arachidonic acid, and 4) sphingolipid-derived ceramide. These pathways have been reviewed elsewhere (14) and will not be discussed further here.

Intracellular Signals Induced by AT_1 Receptors

Ang II promotes its effects by acting directly through Ang II receptors, indirectly through the release of other factors, and via crosstalk with intracellular signalling pathways of other vasoactive agents, growth factors and cytokines. AT_1 receptors are coupled to multiple, specific signalling cascades, leading to diverse biological actions. The signalling processes are multiphasic with distinct temporal characteristics.

AT_1 Signalling Through Phospholipids

Phospholipase C. One of the earliest detectable events resulting from Ang II stimulation is a rapid, PLC-dependent hydrolysis of phosphatidylinositol-4, 5-bisphosphate (15). The PLC family includes three related enzymes: PLC-ß, PLC-g, and PLC-d which are regulated by either G proteins a and ßg (PLC-ß), by tyrosine phosphorylation (PLC-g) or by Ca^{2+} (PLC-d). Classically, AT_1 receptor activation results in a rapid production of 1,4,5-inositol triphosphate (IP_3), and a more sustained release of diacylglycerol, which are involved in Ca^{2+} mobilisation from the sarcoplasmic reticulum and stimulation of protein kinase C (PKC) (16), respectively. Ang II-stimulated IP_3 generation may also be mediated, in part, via tyrosine kinase-dependent pathways.

Increased intracellular Ca^{2+} results in vascular smooth muscle cell (VSMC) contraction, whereas PKC activation regulates intracellular pH through the Na^+/H^+ exchanger (15,16). PLC activation correlates temporally with initiation of contraction in isolated VSMC, as well as in intact small resistance arteries, and most likely constitutes the early signalling pathway for initiation of the Ca^{2+}-dependent, calmodulin activated phosphorylation of the myosin light chain which leads to cellular contraction. The cellular processes underlying Ang II-induced rise in $[Ca^{2+}]_i$ and pHi following AT_1 receptor stimulation have been extensively reviewed elsewhere (15,17).

Phospholipase D. Unlike PLC, which preferentially acts on phospholipids containing phosphoinositol, PLD hydrolyses phosphatidylcholine. The sustained activation of PLD is a major source of prolonged second messenger generation in VSMC and cardiomyocytes (18). Hydrolysis of phosphatidylcholine by PLD leads to the production of phosphatidic acid and subsequent generation of diacylglycerol by phosphatidic acid phosphohydrolase (18). Diacylglycerol is the physiological activator of PKC and is also a source of arachidonic acid. A number of PKC isoforms have been described including a, ß, d, e, μ, V, and l. Molecular mechanisms coupling AT_1 receptors to PLD involve Gßg and their associated Ga12 subunits, Src and RhoA (19). The downstream pathways associated with Ang II-induced activation of PLD in VSMC are PKC independent, but involve intracellular Ca^{2+} mobilisation and Ca^{2+} influx that is tyrosine kinase dependent. Ang II-induced PLD signalling has been implicated in cardiac hypertrophy, VSMC proliferation, and vascular contractility (20,21).

These actions are mediated via phosphatidic acid and other PLD metabolites, that influence vascular generation of superoxide anions by stimulating NAD(P)H oxidase, which activate tyrosine kinases and Raf and modulate intracellular Ca^{2+} signalling (18). The long-term signalling events associated with Ang II-stimulated growth and remodelling in the cardiovascular system are dependent, in large part, on PLD-mediated responses.

Phospholipase A_2. Ang II induces activation of PLA_2, which is responsible for release of arachidonic acid from cell membrane phospholipids (22). Released arachidonic acid is metabolised by cyclooxygenases, lipoxygenases or cytochrome P450 oxygenases to many different eicosanoids in vascular and renal tissues. Cyclooxygenases catalyse the formation of prostaglandin (PG) PGH_2, subsequently converted to thromboxane by thromboxane synthase, to PGI_2 (or prostacyclin) by prostacyclin synthase, or to PGE_2, PGD_2 or PGF_{2a}, by different enzymes (22). Lipoxygenases catalyse the formation of 5-, 12-, or 15-HPETEs, that then undergo spontaneous or peroxidase catalysed reduction to the corresponding HETEs and, in the case of 5-HPETE, to leukotrienes (22). Cytochrome P450 oxygenases catalyse arachidonic acid epoxidation to epoxyeicosatrieenoic acids, w and w-1 hydroxylation to 20- and 19-HETE, and allylic oxidation to other HETEs.

PLA_2-derived eicosanoids influence vascular and renal mechanisms important in blood pressure regulation. In VSMC and endothelial cells, these effects are mediated via AT_1 receptors, whereas in neonatal rat cardiac myocytes, neuronal cells and renal proximal tubule epithelial cells, Ang II-induced activation of PLA_2 occurs via AT_2 receptors (23). Ang II-elicited activation of vascular PLA_2 is dependent on $[Ca^{2+}]_i$, Ca^{2+} calmodulin dependent protein kinase II and mitogen-activated protein kinases (MAPK) (22,23). Activated PLA_2 and its metabolites in turn activate Ras/MAPK-dependent signalling pathways, amplifying PLA_2 activity and releasing additional arachidonic acid by a positive feedback mechanism. Ang II-generated eicosanoids regulate vascular contraction and growth, possibly by activating MAPK and redox-sensitive pathways.

Thromboxanes are involved in Ang II-stimulated contraction, whereas vasorelaxant PGs such as PGE_2 and PGI_2 attenuate Ang II-mediated vasoconstriction in some vascular beds. Lipoxygenase derived eicosanoids also influence Ang II-elicited actions in VSMC. 12-HETE facilitates the stimulatory actions of Ang II on Ca^{2+} transients in cultured cells. Lipoxygenase inhibitors attenuate the vasoconstrictor action of Ang II and decrease blood pressure in SHR (24).

Phospholipid-derived second messengers resulting from Ang II-activated phospholipases. Shown are the major phospholipid-derived products from reactions catalysed by phospholipase A (PLA), phospholipase C (PLC) and phospholipase D (PLD). AT_1R = angiotensin 1 receptor, IP_3 = inositol triphosphate, DAG = diacylglycerol, PKC = protein kinase C.

AT_1-mediated Tyrosine Phosphorylation

A recent development in the field of Ang II signalling is the demonstration that AT_1 receptor activation is associated with increased protein tyrosine phosphorylation and activation of MAPK. These processes are characteristically associated with growth factors and cytokines. Accordingly, it is becoming increasingly evident that in addition to its potent vasoconstrictor properties, Ang II has mitogenic- and inflammatory-like characteristics. Ang II stimulates phosphorylation of many non-receptor tyrosine kinases including PLC-g, Src family kinases, JAK (JAK and TYK), FAK, Ca^{2+}-dependent tyrosine kinases (e.g., Pyk2), p130Cas and phosphatidylinositol 3-kinase (PI3K). In addition Ang II influences activity of receptor tyrosine kinases (RTK), such as EGFR, PDGFR and IGFR. The role of tyrosine kinases in Ang II-mediated signalling has been extensively reviewed (1,2,8). Only recent developments are discussed here.

Non-receptor Tyrosine Kinase Activation

Src family kinases. To date at least 14 Src-related kinases have been identified, of which the 60-kDa c-Src is the prototype. c-Src is abundantly expressed in vascular smooth muscle and is rapidly activated by Ang II in VSMC (25). Src plays an important role in Ang II-induced phosphorylation of PLC-g and IP_3 formation. Src, intracellular Ca^{2+} and PKC regulate Ang II-induced phosphorylation of p130Cas, a signalling molecule involved in integrin mediated cell adhesion.

Src has also been associated with Ang II-induced activation of Pyk2 and extracellular signal-regulated kinases (ERKs) as well as activation of other downstream proteins including pp120, p125Fak, paxillin, JAK2, signal transducers and activators of transcription 1 (STAT1), Ga, caveolin, and the adapter protein, Shc (26). Studies in VSMC isolated from human resistance arteries suggest that c-Src may also be important in the regulation of Ang II-stimulated Ca^{2+} mobilisation (27). Furthermore, c-Src mediates Ang II regulation of plasminogen activity in bovine aortic endothelial cells. Activation of c-Src is required for cytoskeletal reorganisation, focal adhesion formation, cell migration and cell growth (28). Increased activation of c-Src by Ang II may be an important mediator of altered VSMC function in hypertension.

Janus family kinases, tyrosine kinase and STAT activation. Similar to classical cytokine receptors, the AT_1 receptor stimulates JAK2 and TYK2, members of the JAK family . AT_1 receptor-induced activation of JAK leads to phosphorylation of the STAT proteins p91/84 (STAT1a/ß), p113 (STAT2) and p92 (STAT3), which are transcription factors. Ang II-induced tyrosine phosphorylation and nuclear translocation of STAT1 require JAK2 and p59 Fyn kinase (a member of the Src family of kinases). p59 Fyn appears to act as a docking protein for both JAK2 and STAT1, which facilitates JAK2-mediated phosphorylation of STAT1.

JAK proteins are key mediators of mRNA expression and are characterised as "early growth response genes". JAK phosphorylates STAT proteins that are translocated to the nucleus, where they activate gene transcription. Electroporation of antibodies against STAT1 and STAT3 abolished VSMC proliferative responses to Ang II, but not to other growth factors, implicating an essential role of STAT proteins in Ang II-induced cell proliferation. The JAK-STAT signalling pathway activates early growth response genes, and may be a mechanism whereby Ang II influences vascular and cardiac growth, remodelling and repair.

Focal adhesion kinase and proline-rich tyrosine kinase 2. Ang II promotes cell migration and induces changes in cell shape and volume by activating FAK-dependent signalling pathways . Focal adhesion complexes, specialised sites of cell adhesion, act as supramolecular structures for the assembly of signal transduction mediators. The best-characterised tyrosine kinase localised to focal adhesion complexes is a 125-kDa protein, FAK. FAK exhibits extracellular matrix-dependent tyrosine autophosphorylation and physically associates with two non-RTK, c-Src and p59 Fyn (pp59), via their SH2 domains .

FAK autophosphorylation may also result in physical associations with PI3K, which is a 'downstream' tyrosine kinase involved in trophic cellular responses. As a consequence of its association with c-Src, FAK undergoes further tyrosine phosphorylation, which results in FAK binding to Grb2, an association with the GDP-GTP exchange protein, Sos, and Ras. This in turn leads to ERK1/2 activation. FAK is abundant in developing blood vessels, and elevation of its phosphotyrosine content in VSMC is a rapid response to Ang II.

Ang II-induced activation of FAK causes its translocation to sites of focal adhesion with the extracellular matrix and phosphorylation of paxillin and talin, which may be involved in the regulation of cell morphology and movement (32). AT_1-induced FAK activation also plays an important role in Ang II-mediated hypertrophic responses in VSMC. The link between the AT_1 receptor and FAK is unknown, but the Rho family of GTPases may be important.

Another FAK family member, Pyk2, also called cell adhesion kinase-ß, related adhesion focal tyrosinc kinasc and calcium-dependent tyrosine kinase (the rat homologue of Pyk2), is activated by AT_1 receptors and is dependent on increased intracellular Ca^{2+} (32).

Since Pyk2 is a candidate to regulate c-Src and to link G protein-coupled vasoconstrictor receptors with protein tyrosine kinase-mediated contractile, migratory and growth responses, it may be a potential point of convergence between Ca^{2+}-dependent signalling pathways and protein tyrosine kinase pathways in VSMC. In endothelial cells the balance of Pyk2 tyrosine phosphorylation in response to Ang II is controlled by Yes kinase (Src family kinase) and by a tyrosine phosphatase SHP-2 (33).

p130Cas. p130Cas is an Ang II-activated tyrosine kinase that plays a role in cytoskeletal rearrangement (34). This protein serves as an adapter molecule because it contains proline-rich domains, an SH3 domain, and binding motifs for the SH2 domains of Crk and Src. p130Cas is important for integrin-mediated cell adhesion, by recruitment of cytoskeletal signalling molecules such as FAK, paxillin and tensin to the focal adhesions. In cultured VSMC, Ang II induces a transient increase in p130Cas tyrosine phosphorylation (28).

Some investigators have found this phosphorylation to be dependent on Ca^{2+}, c-Src and PKC, and to require an intact cytoskeletal network (28). Other studies reported that Ang II-induced activation of p130Cas is Ca^{2+} and PKC independent (35). Although the exact functional significance of Ang II-induced activation of p130Cas is unclear, it might regulate a-actin expression, cellular proliferation, migration and cell adhesion. p130Cas also plays a critical role in cardiovascular development and actin filament assembly.

Phosphatidylinositol 3-kinase. PI3K is a heterodimeric enzyme composed of a p85 adapter and a p110 catalytic subunit. PI3K catalyses the synthesis of 3-phosphorylated phosphoinositides. The major products of PI3K influence cell survival, metabolism, cytoskeletal reorganisation and membrane trafficking and have recently been identified to play an important role in the regulation of VSMC growth (36). PI3K, characteristically associated with tyrosine kinase receptors, is also activated by AT_1 receptors (37,38). In VSMC Ang II stimulates activity, phosphorylation and migration of PI3K, and induces translocation of the p85 subunit from the perinuclear area to foci throughout the cytoplasm and the cytoskeletal apparatus (37).

PI3K inhibition by wortmannin and LY294002 blocks Ang II-stimulated hyperplasia in cultured rat cells, suggesting the important regulatory role of this non-RTK in VSMC growth (37). Several molecular targets for PI3K have been identified, including centaurin, the actin-binding-protein profilin, phosphoinositide dependent kinases, the atypical PKCs, PLC-g, Rac1, c-Jun N-terminal protein kinase (JNK) and the protein Ser/Thr kinase (Akt)/protein kinase B (PKB) (38). Akt/PKB has recently been identified as an important PI3K downstream target in Ang II-activated VSMC. It regulates protein synthesis by activating p70 S6-kinase and it modulates Ang II-mediated Ca^{2+} responses in aortic cells by stimulating Ca^{2+} channel currents. Akt/PKB has also been implicated to protect VSMC from apoptosis and to promote cell survival by influencing Bcl-2 and c-Myc expression and by inhibiting caspases (38). Mechanisms whereby the AT_1 receptor mediates activation of PI3K-dependent Akt/PKB are unclear, but redox-sensitive pathways and c-Src may be important. Although the exact role of PI3K in Ang II signalling in VSMC has not yet been established, it is possible that this complex pathway may control the balance between mitogenesis and apoptosis.

Tyrosine kinase pathways stimulated by angiotensin II (Ang II) in vascular smooth muscle cells. Ang II activates Src, which regulates phospholipase C-g (PLC-g)- and extracellular signal-regulated kinase (ERK)-dependent signalling pathways. Ang II binding to the angiotensin 1 receptor (AT_1R) induces the physical association and activation of JAK2/TYK2 (Janus kinase/s) as indicated by the dashed line. JAK2/TYK2 phosphorylates STAT proteins (signal transducers and activators of transcription) that are translocated to the nucleus where they activate gene transcription. Ang II also

Receptor Tyrosine Kinases

Increasing evidence suggests that mitogenic responses to AT_1 receptor activation may be mediated by activation of RTK. Ang II can activate RTK, even though it does not directly bind to RTK. This process of transactivation has been demonstrated for EGFR, PDGFR and IGFR and has recently been reviewed . Mechanisms underlying Ang II-induced transactivation of RTKs include activation of tyrosine kinases (Pyk2 and Src) and redox-sensitive processes . EGFR transactivation seems to be a Ca^{2+}-dependent process, whereas PDGFR transactivation is Ca^{2+} independent. Recently, a novel concept for EGFR transactivation by G protein-coupled receptors was suggested.

In response to G protein-coupled receptor agonists (endothelin-1, thrombin, carbachol, lysophosphatidic acid and tetradecanoyl phorbol-13, acetate), heparin binding-EGF is generated by cleavage of pro-heparin-binding-EGF by metalloproteinase (40). Free heparin-binding-EGF then binds to EGFR resulting in EGFR homodimerisation and autophosphorylation. Similar processes have been demonstrated for IGFR transactivation. The role of these mechanisms in RTK transactivation by Ang II is unclear as some studies failed to demonstrate a role for metalloproteinase in AT_1-mediated EGFR transactivation in VSMC and cardiac fibroblasts .

In vitro evidence suggests that AT_1 receptor-induced EGFR transactivation is important for some of the trophic effects of Ang II. For example, AT_1 receptor-elicited tyrosine phosphorylation and activation of EGFR stimulated downstream activation of ERK1/2 and VSMC hyperplasia . In rat VSMC, both Ang II-induced nuclear proto-oncogene expression and increase in c-Fos protein were prevented by treatment with EGFR kinase inhibitor. Ang II-mediated EGFR transactivation also plays a role in p70 ribosomal protein S6-kinase induced protein synthesis. Furthermore, recent studies have demonstrated that EGFR activation is involved in Ang II-induced vascular contraction.

Mitogen-activated Protein Kinases

Mitogen-activated protein kinases are a family of serine/threonine protein kinases that mediate nuclear transduction of extracellular signals by intracellular protein phosphorylation, leading to a cascade of transcription factor activation, enhanced gene expression and trophic cellular responses. Mammalian MAPKs are grouped into six major subfamilies: a) ERK1/2 (also known as p42-kDa MAPK and p44-kDa MAPK, respectively), b) JNK/stress-activated protein kinases (JNK/SAPK), c) p38 MAPK, d) ERK6, p38-like MAPK, e) ERK3, and f) ERK5 (also called Big MAPK 1) . MAPK-dependent signalling pathways have been associated with cellular growth and apoptosis, cellular differentiation and transformation and vascular contraction. ERK1/2 is activated in response to growth and differentiation factors, whereas JNKs and p38 MAPK are usually activated in response to inflammatory cytokines and cellular stress .

Ang II differentially activates the three major members of the MAPK family, ERK1/2, JNKs and p38 MAPK (43,44). Induction of MAPK activation typically involves phosphorylation by a MAPK kinase, also known as MEK. MEK is, in turn, regulated by other MEK kinases, including Raf-1. Although activated by similar stimuli, the signalling processes leading to JNK and p38 MAPK activation are quite different. The best characterised MAPK cascade is the Raf-Ras-MEK-ERK1/2 pathway.

Events downstream to MAPK activation are numerous and heterogeneous and include PLA_2, cytoskeletal proteins, the MAPK-activated protein kinase 2, and the $pp90^{rsk}$ protein kinase, which can translocate to the nucleus and activate transcription factors . Once phosphorylated, ERKs translocate to the nucleus to phosphorylate transcription factors and thereby regulate gene expression of cell cycle-related proteins. In VSMC, another downstream target of ERK is the serine/threonine protein kinase $pp90^{rsk}$, which phosphorylates the S6 ribosomal protein and stimulates protein synthesis. ERK1/2 activation ultimately results in enhanced proto-oncogene expression, and activation of the activator protein 1 complex transcription factor and probably regulates cell cycle progression as well as protein synthesis in VSMC. Ang II may also induce protein synthesis by an ERK-independent pathway in part via activation of the 70-kDa S6-kinase. Other downstream targets of MAPK include cyclooxygenase-2, the contractile regulatory protein h-caldesmon, the high-molecular weight form of caldesmon, myelin basic protein, microtubule-associated protein, Ca^{2+} channels and the Na^+/H^+ exchanger . The functional outcome of MAPK activation probably depends in part on the availability of downstream substrates.

Ang II activates the MAPK signalling cascade at various intracellular levels. It induces tyrosine and threonine phosphorylation of ERK1/2, JNK/SAPK and p38 MAPK in cultured VSMC, as well as in intact arteries. It stimulates phosphorylation of Ras, Raf and Shc and it increases activity of MEK kinase and MEK. In addition, Ang II increases activation of vascular Src and Pyk2, potential links between the Ang receptor and ERK. Ang II-stimulated ERK1/2 is associated with increased expression of the early response genes c-*fos*, c-*myc* and c-*jun*, DNA synthesis, cell growth and differentiation and cytoskeletal organisation .

In addition to ERKs, Ang II activates JNK/SAPKs, which regulate VSMC growth by promoting apoptosis or by inhibiting growth . Ang II induces phosphorylation of JNK/SAPK via p21 activated kinase (aPAK) which is dependent on intracellular Ca^{2+} mobilisation and on PKC activation. Following phosphorylation, the isoforms JNK-1 and JNK-2 translocate to the nucleus to activate transcription factors, such as c-Jun, ATF-2 and Elk-1 (46). Ang II appears to activate VSMC ERK1/2 and JNK/SAPK via different signalling pathways. ERK phosphorylation occurs via a Ca^{2+}-dependent or -independent pathway that involves c-Src and the atypical PKC isoform PKC-x, whereas JNK/SAPK activation occurs via a Ca^{2+}-dependent pathway that involves a tyrosine kinase other than Src and a novel PKC isoform . Furthermore, whereas Ang II-induced phosphorylation peaks within 5 min, kinase activation is maximal at about 30 min. The functional effects of Ang II-induced signalling of ERK1/2 and JNK/SAPK in VSMC probably relate to regulation of cell growth. Ang II-activated ERK1/2 and JNK/SAPKs have opposite growth effects, with ERK1/2 being facilitative and JNK/SAPK inhibitory. These signalling processes and associated cellular functions are important in vascular damage associated with cardiovascular disease.

Ang II also phosphorylates vascular p38 MAPK, which plays an important role in inflammatory responses, apoptosis and inhibition of cell growth (48). The p38 MAPK pathway has been implicated in various pathological conditions including cardiac ischemia, ischemia/ reperfusion injury, cardiac hypertrophy, progression of atherosclerosis and arterial remodelling in hypertension.

The specific upstream and downstream regulators of Ang II-activated p38 in VSMC are unclear, but p38 MAPK could be a negative regulator of ERK1/2 (48). p38 MAPK has been implicated to be an essential component of the redox-sensitive signalling pathways in Ang II-activated VSMC (48).

Inactivation of Ang II-stimulated MAPK occurs via MAPK phosphatase 1 (MKP-1) induced dephosphorylation of both tyrosine and threonine on MAPK. Inhibition of MKP-1 results in sustained activation of MAPK in response to Ang II, suggesting that this enzyme is primarily responsible for the termination of the MAPK signal (49).

Ang II induces activation of MKP-1, as well as tyrosine phosphatase (PTP-1C), and Ser/Thr phosphatase PP2A (49). These effects appear to be mediated via the AT_2 receptor subtype, which has been associated with inhibition of cell growth and apoptosis. Accordingly, AT_1 receptors induce growth via stimulation of ERK-dependent signalling pathways, whereas AT_2 receptors oppose these effects by stimulating MKP-1 activity to inhibit ERK activity, and to arrest the cell growth signal. Termination of Ang II-stimulated MAPK activity may also involve activation of PKA, which inhibits the phosphorylation of Raf-1.

Ang II-stimulated extracellular signal-regulated kinase (ERK)-dependent signalling pathways in vascular smooth muscle cells. The ERK phosphorylation cascade is initiated by Ang II binding to angiotensin 1 receptors (AT_1R) that induce Shc-Grb2-Sos formation (tyrosine phosphorylation of Shc) which in turn facilitates guanine nucleotide exchange on the small G protein Ras-GDP/GTP.

Activated Ras-GTP interacts with the Ser/Thr kinase Raf (mitogen-activated protein kinase (MAPK) kinase kinase (MAPKKK)) which translocates to the cell membrane. Activation of Raf leads to phosphorylation of two serine residues present in MEK (MAPK/ERK kinase) which in turn phosphorylates Thr/Tyr and activates MAPK, present as a 44- (ERK-1) and a 42-kDa isoform (ERK-2). Phosphorylated ERK has diverse intracellular protein targets which it phosphorylates and activates. Dephosphorylation of ERK is accomplished by activation of MAPK phosphatase-1 (MKP-1). PLA_2 - phospholipase A_2.

Small G Proteins (Rho family)

In addition to signalling through heterotrimeric G proteins, recent evidence suggests that AT_1 receptors activate monomeric small (21 kDa) guanine nucleotide-binding proteins (small G proteins) in VSMC. The small G protein superfamily comprises five subfamilies (Ras, Rho, ADP ribosylation factors, Rab and Ran) that act as molecular switches to regulate cellular responses (50). Of these, the Rho subfamily (RhoA, Rac1 and Cdc42) has been associated with Ang II signalling. In its GDP-bound form, RhoA is inactive and is mainly cytoplasmic. RhoA is activated upon the exchange of bound GDP for GTP and for its activation, requires post-translational modification (geranyl-geranylation) (50). One of the major downstream targets of RhoA is Rho kinase, which promotes VSMC contraction via the phosphorylation of the myosin-binding subunit of myosin light chain phosphatase, thereby inhibiting phosphatase activity (51). This contributes to increased Ca^{2+} sensitisation. Various vasoactive agonists signalling through G protein-coupled receptors enhance the sensitivity of the contractile machinery to changes in $[Ca^{2+}]_i$ via RhoA/Rho kinase. Whether this system contributes to contraction following AT_1 receptor activation awaits clarification. However, RhoA seems to play a role in AT_1-stimulated growth signalling pathways and in processes associated with atherosclerosis. Yamakawa et al. (52) demonstrated in rat VSMC that Ang II increases c-fos expression and protein synthesis via RhoA/Rho kinase-dependent mechanisms that do not involve ERK1/2 or p70 S6-kinase activation. Funakoshi et al. (53) reported that Rho kinase mediates Ang II-induced monocyte chemoattractant protein-1 expression in rat VSMC. RhoA has also been shown to be important in AT_1-mediated PLD activation.

Ang II also activates Rac1, another small G protein. Rac1 participates in cytoskeletal organisation, cell growth, inflammation and regulation of NAD(P)H oxidase (50). In VSMC, Ang II activates Rac1, which is an upstream regulator of p21-activated kinase and JNK. Rac1 also plays a role in Ang II-induction of gene transcription and in the regulation of NAD(P)H oxidase, which mediates generation of superoxide anions ($.O_2^-$) in vascular cells (50).

Rho-regulated signalling pathways by angiotensin II (Ang II). Rho is activated through AT_1 receptors. Rho regulates the activity of many signalling pathways that influence contraction, growth and cytoskeletal organisation. MBS, myosin binding subunit of the myosin light-chain phosphatase; PRK2, protein kinase C-related kinase; SRF, serum response factor

Generation of Reactive Oxygen Species

Reactive oxygen species such as O_2^- and hydrogen peroxide (H_2O_2) act as intercellular and intracellular second messengers that may play a physiological role in vascular tone and cell growth, and a pathophysiological role in inflammation, ischemia-reperfusion, hypertension and atherosclerosis. Xanthine oxidase, mitochondrial oxidases and arachidonic acid are the major sources of oxidative molecules in non-vascular tissue, whereas a nonmitochondrial, membrane-associated NAD(P)H oxidase appears to be the most important source of $.O_2^-$ in vascular cells (54). This enzyme transfers electrons from NADH or NADPH to molecular oxygen, producing $.O_2^-$. The complete molecular structure of the vascular oxidase is unknown, but it shares some features with the neutrophil oxidase, which comprises five subunits: a 22-kDa a-subunit ($p22^{phox}$), a glycosylated 91-kDa ß-subunit ($gp91^{phox}$), which together make up cytochrome b_{558}, the electron transfer element; cytosolic components $p47^{phox}$ and $p67^{phox}$, and a low-molecular weight G protein, Rac1 or Rac2. Recent studies by Rueckschloss et al. (55) demonstrated complete homology of the cDNA sequence of $gp91^{phox}$ in human umbilical vein endothelial cells and neutrophils. Alternatively, differences in the levels of gene expression of $gp91^{phox}$ and $p67^{phox}$ were observed, with human umbilical vein endothelial cells exhibiting 1.1 and 2.5%, respectively, of the levels of expression of these genes, compared with those of neutrophils. By contrast, gene expression for $p22^{phox}$ and $p67^{phox}$ was comparable between these two cell types. These observations suggest that the levels of expression of $gp91^{phox}$ and $p67^{phox}$ genes may be rate-limiting factors for human vascular cell $.O_2^-$ production. Upon activation, the $p47^{phox}$ and $p67^{phox}$ proteins are translocated to the membrane and associate with the cytochrome b_{558}, creating the active oxidase. In VSMC, Ang II increases $.O_2^-$ production by activating NAD(P)H oxidase. This effect is sustained and probably contributes to long-term signalling events such as cell growth.

Generation of reactive oxygen species is regulated by various cytokines and growth factors, including Ang II, which increases $.O_2^-$ and H_2O_2 production in cardiac, vascular smooth muscle, endothelial, adventitial and mesangial cells (21,54,56). Increased production of reactive oxygen species has been implicated in the pathogenesis of Ang II-induced but not catecholamine-induced hypertension (57). Mechanisms underlying oxidative stress-induced hypertension may be associated with the vascular mitogenic effects of $.O_2^-$ and H_2O_2, decreased bioavailability of endothelium-derived nitric oxide and contractile actions of $.O_2^-$ and H_2O_2. Growth of VSMC by Ang II has an essential redox-sensitive component, which appears to be mediated in part via activation of p38 MAPK and JNK (48). Another redox-sensitive process cascade whereby Ang II influences cell function is through phosphorylation of the cell survival protein kinase Akt/PKB and activation of NF-kB, important in inflammatory responses (54).

Generation of reactive oxygen species in the vasculature. Many enzyme systems stimulate production of superoxide anion ($\cdot O_2^-$) from O_2. These include NADH/NADPH oxidase, xanthine oxidase, lipoxygenase, cyclooxygenase, P450 mono-oxygenase and mitochondrial oxidative phosphorylation. NADH/NADPH oxidase is a multisubunit enzyme that is the major regulated source of reactive oxygen species in endothelial and vascular smooth muscle cells. Dismutation of $\cdot O_2^-$ spontaneously or enzymatically by superoxide dismutase (SOD) produces hydrogen peroxide (H_2O_2) that can undergo further reactions to generate the highly reactive hydroxyl radical ($\cdot$OH). H_2O_2 may be metabolised by catalase or peroxidases to H_2O and O_2. Downstream targets of $\cdot O_2^-$ and H_2O_2 include ERK5, p38, tyrosine kinases and NF-kB. MAPK = mitogen-activated protein kinase; ERK = extracellular signal-regulated kinase; Akt/PKB = protein Ser/Thr kinase

Ang II-induced Expression of Proto-oncogenes and Growth Factors

Long-term control of Ang II-regulated cellular growth, adhesion, migration, fibrosis and collagen deposition within the vasculature involves protein synthesis. Ang II induces the expression of several proto-oncogenes in human and rat VSMC, including c-*fos*, c-*jun*, c-*myc*, *erg-1*, VL-30, and proto-oncogene/activator protein 1 complex (58). Stimulation of early response genes by Ang II is associated with increased gene expression and production of growth factors, such as PDGF, EGF, transforming growth factor-ß, IGF-1, basic fibroblast growth factor and platelet activating factor, vasoconstrictor agents, such as ET-1, adhesion molecules such as ICAM-1, VCAM-1 and E-selectin, and integrins anß3 and ß5, and chemotactic factors such as tumour necrosis factor-a and monocyte chemoattractant protein-1 (58). These agents may contribute indirectly to the trophic and inflammatory actions of Ang II in cardiovascular tissue.

Ang II influences the architecture and integrity of the vascular wall by modulating cell growth as well as regulating extracellular matrix composition. Ang II increases expression and production of fibronectin, collagen type 1, tenascin, glycosaminoglycans, chondroitin/dermatan sulfates and proteoglycans, major constituents of the extracellular matrix in the vessel wall. In VSMC, mesangial cells and endothelial cells, Ang II increases levels and activity of plasminogen activator inhibitor-1, influencing fibrinolysis, extracellular matrix turnover and degradation and regulation of cell migration. Some of these actions have been linked to the AT_4 receptor subtype. However, this remains to be clarified. Ang II also stimulates the activity of matrix metalloproteinases (58) responsible for extracellular matrix degradation. Accordingly, Ang II influences vascular structure by stimulating synthesis of structural components of the extracellular matrix and by increasing production of factors that degrade the extracellular matrix proteins.

Conclusions

Ang II activates multiple signalling pathways. Until recently, the signalling events elicited by Ang II were considered to be rapid, short-lived and divided into separate linear pathways. However, these major intracellular signalling cascades do not function independently and are actively engaged in crosstalk. Downstream signals from the Ang II-bound receptors converge to elicit complex and multiple responses (reviewed in 59,60). Ang II induces a multitude of actions in various tissues, and the signalling events following occupancy and activation of angiotensin receptors are tightly controlled and extremely complex. Furthermore, the fact that Ang II transactivates EGFR, PDGF and IGFR suggests that RTKs are downstream targets of AT_1 receptors. Alterations of these highly regulated signalling pathways in cardiovascular cells may be pivotal in structural and functional abnormalities that underlie vascular pathological processes such as hypertension and atherosclerosis. Although there has been significant progress in the last few years in the elucidation of aberrations in Ang II-induced signal transduction in hypertension, we still know very little about the processes that underlie these phenomena, and at what point some pathways become more important than others. With the availability today of molecular and pharmacological tools that allow manipulation of specific signalling molecules, identification of distinct abnormalities in intracellular signalling should be possible. This will further our understanding of the role of Ang II in physiological and pathophysiological processes and will help define novel therapeutic targets in cardiovascular disease.

Improving Human Life, Beginning with the Basics

If we can master the fundamental biological processes that shape life, we will possess the basic information needed to address diseases that afflict individuals at every stage of their lives. The University's initiative, "The Basics of Life," expressly capitalises on this concept. An infusion of private funds will catalyse the investment of others, including government and industry, and enable us to build on our leadership in key areas of basic and clinical research, accelerating the development of lifesaving therapies. In the process, we will establish a powerful model for research universities across the nation.

Creating Knowledge To Save Lives

The premise behind the University of Virginia's initiative, The Basics of Life, is compelling. If we can understand fundamental issues—such as how cells become specialised and form tissue, how they communicate with each other and respond to their environment, and how they migrate through the body—we will gain the insight needed to cure a panoply of diseases that affect human beings.

When lives are in the balance, basic information is crucial. Only through detailed knowledge of the formation, maintenance, degeneration, and regeneration of molecular networks, tissues, and organs can we prevent birth defects; control the rampant tissue growth associated with cancer; slow the deterioration that characterises aging; and facilitate the repair, regeneration, and replacement of injured tissues. Fundamental advances in biological research promote the development of diagnostic tools and therapies throughout the entire life span. Breakthroughs at any one stage of human development can affect treatment at every other stage. At U.Va., we propose to reorganise our research efforts in new and creative ways so that we can more effectively apply our Basics-of-Life strategy to cure or prevent disease, while developing collaborations and partnerships worldwide to extend this model to other major research universities. We will create University-wide initiatives through three institutes and a research centre: the Morphogenesis and Regenerative Institute, the Virginia Institute for Clinical and Translational Research, the Institute on Aging, and the Pediatrics and Birth Defects Research Centre.

Building from a Position of Strength

Thanks to strategic investments by the University and the Commonwealth during the last five years and the extraordinary success of our faculty in securing outside research funding, we have assembled many of the elements needed to assume national leadership in driving lifesaving breakthroughs in diagnosis and treatment of cancer, birth defects, diabetes, and heart disease through advanced research in biology and medicine.

Morphogenesis and Regenerative Medicine Institute

Founded in 2002, the multidisciplinary Morphogenesis and Regenerative Medicine Institute was formed explicitly to pursue one of the most promising lines of inquiry in biology and medicine: expanding our knowledge of how cells differentiate and organise themselves so that we can harness these insights to regenerate defective or damaged organs. The potential of discoveries in this area is stunning, offering us the ability to repair spinal cord injuries, for instance, or restore lung tissue damaged by years of smoking or coal mining.

The institute brings together U.Va.'s world-leading programs in such fundamental biological processes as the migration of cells through the body, the mechanisms that enable cells to adhere to each other, and the signalling pathways that control the organisation of tissues. It also draws on our outstanding expertise in kidney and blood vessel development, among other areas. Virginia Institute for Clinical and Translational Research.

The newly created Virginia Institute for Clinical and Translational Research will play a critical role in extending the benefits of basics-of-life science to the diseases of adults. This research "translates" basic scientific discoveries into what Mr. Jefferson called "useful science" that has direct applications in patient care. Initially, we intend to combine our expertise in tissue development and diabetes to transform precursor cells in the pancreas to cells that can express insulin. The result of this dramatic breakthrough: a cure for Type 1 diabetes. In the process, we will extend our expertise to other areas, such as vascular disease and cancer, which can benefit from similar approaches.

Institute on Aging

The University's Institute on Aging brings together leading authorities in psychology, biology, medicine, and engineering from the United States and Europe to focus on mobility and independence, cognitive aging, age-related neurological disease, and other important issues. U.Va. scientists are responsible for one of the world's foremost longitudinal studies of aging, allowing medical researchers to compare declines in cognitive function with changes at the cellular and molecular level. These correlations provide the fundamental insights needed to pursue more effective strategies to bolster cognitive function and to prevent and treat such a Ge-related illnesses as Parkinson's and Alzheimer's.

Taking the Next Step

The University has put in place many of the elements needed to realise the promise of its basics of life strategy, and the University's Board of Visitors has committed to strengthen scientific and medical research even further. In recent months, the University has recruited several distinguished senior researchers in neuroscience, cell biology, and epidemiology, all areas that are vital to our knowledge of cell, tissue, and organ development. To complement the exciting work of these institutes, we will establish a Pediatrics and Birthdefects Research Centre that combines

University expertise in the Schools of Medicine, Education, and Law. The Pediatrics and Birth Defects Research Centre will apply insights from basic research to address major threats to the health of the fetus, optimise the care of the newborn, and address the treatment of children with acute brain injury or spinal cord injury. The work of the centre will gain additional momentum from the construction of a new U.Va. Children's Hospital.

There is, nonetheless, much more work to be done. An infusion of significant private funding will jump-start our efforts ($300 million in endowment will be required to support properly the four world-class centres). This support will enable us to make strategic additions to our faculty, develop advanced research instruments and facilities, and assemble multidisciplinary research groups that can speed the translation of basic research into real-world clinical applications. With these new resources, the University of Virginia can become the place where birth defects are prevented, where we learn how to prevent cancer cells from moving to lymphatic nodes, where both Type 1 and Type 2 diabetes are cured, where early heart attacks are prevented, and where cognitive and physical functionality are sustained in old age.

In short, by enabling us to complete and energize this unique network devoted to biology and medicine over the course of the human life, these additional funds will accelerate the flow of lifesaving therapies to those who need them most. Our goal? To improve human life, beginning with the basics.

Molecular Biology

Genome sequence comparison has been an important method for understanding gene function and genome evolution since the early days of gene sequencing. Alignment of DNA sequences is the core process in comparative genomics. In recent years, an important new sequence-analysis task has emerged: comparing an entire genome with another. Several powerful alignment algorithms have been developed to align two or more sequences.

MUMmer

MUMmer is a system for rapidly aligning entire genomes, whether in complete or draft form. MUMmer can also align incomplete genomes; it can handle thousands of contigs from a shotgun sequencing project, and will align them to another set of contigs or a genome using the NUCmer program included within the system. If the species are too divergent for a DNA sequence alignment to detect similarity, then the PROmer program within the environment can generate alignments based upon the six-frame translations of both input sequences. The original MUMmer system, version 1.0, was described in a 1999 Nucleic Acids Research paper. Version 2.1 appeared a few years later and was described in a 2002 Nucleic Acids Research paper, and the most recent version MUMmer 3.0 was described in a 2004 Genome Biology paper.

BLAT

BLAT (The BLAST-Like Alignment Tool) is a new tool for sequence alignment, which is similar in many ways to BLAST. The program rapidly scans for relatively short matches (hits), and extends these into high-scoring pairs (HSPs). However, BLAT differs from BLAST in several significant ways. Specifically, where BLAST builds an index of the query sequence and then scans linearly through the database, BLAT builds an index of the database and then scans linearly through the query sequence. Where BLAST triggers an extension when one or two hits occur in proximity to each other, BLAT can trigger extensions on any number of perfect or near-perfect hits. Where BLAST returns each area of homology between two sequences as separate alignments, BLAT stitches them together into a larger alignment. Both the client/server and the stand-alone can do comparisons at the nucleotide, protein, or translated nucleotide level.

MEGABlast

Mega BLAST uses the greedy algorithm of Zhang et al. for nucleotide sequence alignment search and concatenates many queries to save time spent scanning the database. This program is optimised for aligning sequences that differ slightly as a result of sequencing or other similar "errors". It is up to 10 times faster than more common sequence similarity search and alignment programs and therefore can be used to swiftly compare Microbial population genetics is a rapidly advancing field of investigation with relevance to many areas of science. The subject encompasses theoretical issues such as the origins and evolution of species, sex and recombination. Population genetics lays the foundations for tracking the origin and evolution of antibiotic resistance and deadly infectious pathogens and is also an essential tool in the utilisation of beneficial microbes.

Written by leading researchers in the field, this invaluable book details the major current advances in microbial population genetics and genomics. Distinguished international scientists introduce fundamental concepts, describe genetic tools and comprehensively review recent data from SNP surveys, whole-genome DNA sequences and microarray hybridisations. Chapters cover broad groups of microorganisms including viruses, bacteria, archaea, fungi, protozoa and algae. A major focus of the book is the application of molecular tools in the study of genetic variation. Topics covered include microbial systematics, comparative microbial genomics, horizontal gene transfer, pathogenic bacteria, nitrogen-fixing bacteria, cyanobacteria, microalgae, fungi, malaria parasites, viral pathogens and metagenomics.

An essential volume for everyone interested in population genetics and highly recommended reading for all microbiologists.

The higher taxonomic groups within prokaryotes are presently distinguished mainly on the basis of their branching in phylogenetic trees. In most cases, no molecular, biochemical or physiological characteristics are known that are uniquely shared by species from these groups. Analyses of genome sequences are leading to discovery of novel molecular characteristics that are specific for different groups of bacteria and archaea and provide more precise means for identifying and circumscribing these groups of microbes in clear molecular terms and for understanding their evolution. These new approaches and their limited applications for clarifying microbial systematics are described here. Because of their taxa specificities, further studies on these newly discovered molecular characteristics should lead to discovery of novel biochemical and physiological characteristics that are unique to different groups of microbes.

Microbes are ubiquitous in the world in which we live. With the development of high throughput DNA sequencing technology, there has been an explosion of DNA sequence data on microbes. The major aim of future microbial genomics will be to identify the functional significances of individual gene and genomic fragments and to use the information to help improve human health and promote our society development. One current major undertaking to understand genomic information is the comparative analyses between genomes that are not only distantly related, but also closely related ones.

Such comparative analyses between genomes that have diverged at different evolutionary time scales allow us to extract different types of information about biological functions and evolutionary processes. We review the tools and databases that have been established for comparative analyses of microbial genomes and discuss the implications of such analyses on our understandings of the common properties of life, the extent of genome plasticity and diversity within and between species, the processes and mechanisms underlie the observed genome diversity, and the origin and evolution of life.

Horizontal gene transfer, as a major force in shaping bacterial gene content, has gained incredible attention over the last decade. Along with the fast growing bacterial genome sequence data, there have been an increasingly large number of studies focused on horizontal gene transfer. The studies have been gradually transformed from identifying individual genes that have been horizontally transferred to assessing the general patterns of horizontal gene transfer and evaluating the systematic consequences of massive gene transfers.

The rates of gene transfers have been measured by various methods such as parsimony and maximum likelihood methods. Different phylogenetic methods were applied to a variety of data sets to assess whether there exists a congruent and meaningful bacterial tree. Even though some consensus has been reached, many contradictions have emerged and need to be solved in future studies. Our goal here is to review recent studies on this subject with an emphasis on the emerging patterns in horizontal gene transfer research.

Population genetics examine variation in genes among a group of strains of a particular species. Its major theme is to look at how different environmental factors and selective pressures can affect the distribution of genes and alleles. In this chapter *Yersinia pestis* was employed as an example to illustrate how the techniques are used for population genetic studies and how the achievements of these kinds of studies can be used for rapid identification and tracing the origin of pathogenic bacteria. New emerging techniques, including high throughput sequencing technologies, will give us unprecedented opportunities to understand microevolution and pathogenesis of bacterial pathogens.

Symbiotic nitrogen-fixing Rhizobia are of global significance, both in terms of their ecological relationships and their importance as an environmentally benign source of nitrogen for crop plants. These bacteria are capable of forming mutualistic relationships with a variety of legume hosts, where they convert atmospheric nitrogen (N_2) to ammonia that is used to help meet the nitrogen needs of the host plant. In this chapter, we review our current understanding of the population genetics of this diverse group of bacteria. First we briefly describe the various types of genomic architectures that are found within rhizobial species. Next we outline the phylogenetic relationships among the recognised taxa. We then summarise the results of studies that have examined patterns of molecular genetic variation in rhizobial populations at local, regional, and global levels.

The results of these studies have revealed several important insights, including the existence of extensive diversity within, as well as significant genetic differentiation between local and regional populations. The results also provide evidence for long-distance gene flow between continental populations. Several studies have also indicated the existence of low-to-intermediate levels of recombination within rhizobial populations. Fine-scale studies of specific genomic components (e.g., symbiotic plasmids) have also shown that certain genomic elements appear to be more prone to recombination than others.

The results also showed that the different loci responsible for the development of the symbiosis appear to be under different forms of selection. Because most the population studies to date have focused on strains from root nodules, surprisingly little is known on the population genetics of the more numerous non-nodulating soil rhizobia. Future efforts to characterise these populations should significantly enhance our ability to manipulate rhizobial populations in agricultural ecosystems.

Cyanobacteria are a group of ecologically diverse photosynthetic bacteria. Because niche differentiation is ultimately the product of differences among individuals within populations, understanding the evolutionary origins of this diversity ultimately requires a population genetics perspective. This chapter considers the current state of our understanding of the mechanisms that generate variation in cyanobacteria, the distribution of this diversity and its potential functional importance, with an emphasis on recent work from our laboratory regarding diversity within and among populations of the cosmopolitan, multicellular cyanobacterium, *Mastigocladus laminosus*. It concludes with an appraisal of the potential to take a population genomics approach to address fundamental questions regarding the nature of adaptive variation and niche differentiation in these microorganisms.

Algae are a highly diverse group of protists, ranging from simple, unicellular organisms to complex, multicellular entities with a range of differentiated tissues and distinct organs. They are found among diverse aquatic ecosystems and play important roles by supplying carbon and energy as well as providing habitat to other members of the biological communities. Some algae cause significant environmental and health problems. However, despite their importance, relatively little is known about this group of organisms.

The first section of this chapter briefly summarises our current understanding of the diversity and evolution of the algae. The second section reviews the molecular markers used to address algal population genetic issues. The third section summarises the population genetic analyses of three algal groups: the dinoflagellates, the diatoms and the haptophytes. With the application of genomics information and technology, the field of algal population genetics is approaching an exciting period of development and expansion.

Mutualisms are reciprocal exploitations that nonetheless increase the fitness of each interacting partner. Two groups of fungal mutualists of plants, epichloë endophytes of grasses and arbuscular mycorrhizal fungi, were selected as focal systems to discuss population-level processes that contribute to the establishment and maintenance of mutualistic interactions. These two classes of fungal cooperators of plants are subject to different and often conflicting selective pressures and represent distinct trajectories of mutualism evolution. Yet, in both cases population structure of symbionts is a source of information critical for understanding how these fungi interact with their hosts.

This article reviews the more common DNA-based markers that are used to genotype fungi, as well as other eukaryotes, and presents several examples of their application to elucidate the population genetics of mammalian pathogenic fungi. The most common fungal infections are briefly summarised to illustrate the issues that are routinely addressed by population genetics approaches. This exciting area of research is continually growing as the methods become more accessible and medical mycologists recognise the value of studying natural isolates of pathogenic fungi.

Genetic diversity, population structure, and evolutionary history of malaria parasites are among the key factors that will influence our ability to identify genes contributing to drug resistance, parasite development, and disease pathogenesis. These factors also have an impact on vaccine and drug development, parasite source tracking, as well as the formulation of other disease prevention and control measures. For example, a highly polymorphic parasite population will contain ample genetic diversity capable of generating drug resistance genotypes at an accelerated rate; while the presence of homogeneous parasite populations should aid in the development of an effective malaria vaccine. Malaria research in the post-genomic era offers many new tools for use in population genetics analyses.

Many viral pathogens, especially those with an RNA genome, are characterised by their high mutation rates and large population sizes. These features are responsible for the high levels of genetic variation usually found in viral populations and for their rapid response to different selective challenges encountered during their infection and transmission processes. They are quantitatively and qualitatively so different from most other organisms that special models and concepts, such as the quasispecies model, have been developed to better describe the evolutionary dynamics of viral populations. Here, we review these and other salient features of viral pathogens, with a species emphasis on RNA viruses. We describe how population genetics theory provides an adequate framework for analysing and interpreting genetic variation in viral populations, for understanding their dynamics, and to study adaptive processes occurring therein. However, not all the evolutionary changes observed are due to the action of positive selection and this is not an all-mighty agent of evolutionary change. We finish this review by introducing a recently developed framework for integrating the evolutionary and epidemic behaviour of infectious organisms, known as phylodynamics, which is especially well-suited for fast evolving organisms such as viruses.

Microbial population genetics examines the spatial and temporal patterns of genetic variation across diverse geographic scales and ecological niches. With the arrival of molecular biological techniques, the past 40 years have seen tremendous progresses in microbial population genetics. However, in recent years, the analyses of genetic materials directly from natural environments have revolutionised our approaches and understandings of the diversity, function, and inter-relationships among microorganisms in diverse natural ecological niches. The emergence and development of this expanding new field, that of metragenomics, has been primarily driven by technical and analytical methods developed from high throughput platforms for cloning, microfluidics, DNA sequencing, robotics, high-density microarrays, 2D-gel electrophoresis, and mass spectrometry as well as associated bioinformatics softwares.

In this chapter, I present a general overview of this exciting development, including the general approaches and common tools used for identifying the diversity and function of microorganisms in natural biological communities. Of special notes are the potential impacts of recent developments in single cell isolation, whole-genome amplification, pyrosequencing, and database warehousing on our understanding of microbial population structures in nature. These exciting developments are bringing significant opportunities as well as new challenges to the field of microbial population genetics.

Epigenetics

Epigenetics is the study of changes in gene expression caused by mechanisms other than changes in the DNA sequence. Epigenetics is a rapidly advancing field with an increasing impact on biological and medical research.

The editors of this book have assembled top-quality scientists from diverse fields of epigenetics to produce a major new volume. Comprehensive and cutting-edge, the 26 chapters in this book constitute a key reference manual for everyone involved in epigenetics, DNA methylation, cancer epigenetics and related fields.

Topics include: early life environment, DNA methylation and behaviour, histone acetyltransferase biology, transgenerational epigenetic inheritance, mammalian X inactivation, epigenetic memory in plants, polycomb group regulation, centromeres and telomeres, DNA sequence contribution to nucleosome distribution, macrosatellite epigenetics, histones, cell-fate specification and reprogramming, DNA methylation in cancer, variant histone H2A and cancer development, RNA modification, paramutation in plants, DNMT3L dependent methylation during gametogenesis, non-coding RNA, bisulphite-enabled technologies, rapid analysis of DNA methylation, microarray mapping, DNA methylation profiling, ChIP-sequencing, genome-wide DNA methylation analysis, and epigenetics in maize. In addition there are useful chapters on bioinformatics in epigenomics, resources and tools for epigeneticists, and educational resources for epigenetics.

This up-to-date reference manual is an essential book for those working in the field and for scientists in other disciplines it represents a major information resource on the fascinating and fast-moving field of epigenetics.

The DNA molecule contains within its chemical structure two layers of information. The DNA sequence that bears the ancestral genetic information and the pattern of distribution of covalently bound methyl groups to cytosines in DNA. While the genetic information is similar in all tissues in the individual, the pattern of distribution of methylation across the genome is cell-type specific. DNA methylation is an important regulator of gene function. Recent data that will be discussed here that supports the hypothesis that DNA methylation is a reversible biological signal.

This expands the potential role of DNA methylation beyond embryogenesis to other time-points in life and to post mitotic tissues such as the brain. DNA methylation is proposed to act as a genomic response to both physical and social signals from the environment at different time points in life and to serve as a genomic memory of these exposures at different time scales, stably altering gene expression programming and thus modulating the physical and behavioural phenotypes to respond to these environments. It is hypothesized that DNA methylation provides within the structure of the DNA a dynamic interface between the changing world around us and the relatively fixed and stable genome.

A histone $(H3\text{-}H4)_2$ tetramer flanked by two H2A-H2B heterodimers form the core protein structure, around which DNA is wrapped. DNA and the histone octamer together form the smallest chromatin particle, the nucleosome. How intimately the DNA associates with the core histones and how tightly the nucleosomes are packed with each other is determined by a key post-translational modification of the histone proteins, namely acetylation. Histone acetylation was first discovered in the early 1960s.

After a dearth of progress, due to technical limitations, our knowledge of histone acetylation has exploded in the last fifteen years. Enzymes that catalyse acetylation of histones, the histone acetyltransferases, have been discovered, proteins associated with these have been identified and their preferences for specific histone residues have been determined. Importantly, we are gaining a better understanding of the relevance of histone acetylation in health and disease through the discovery of genetic mutations underlying human diseases in loci encoding histone acetyltransferases (HATs) and through examination of mouse strains deficient in specific histone acetyltransferases. Here we discuss the principles of histone acetyltransferase biology.

Epigenetic states are faithfully inherited through mitotic cell division, but are generally cleared and reset on passage through the mammalian germline. But this clearing of epigenetic marks is not always complete, leading to transgenerational inheritance of epigenotype. Transgenerational epigenetic inheritance has been demonstrated in several organisms, including mammals, and has been most comprehensively studied in mouse strains carrying variants of the *agouti* (A^{vy}) and *axin* ($Axin^{Fu}$) alleles. The most prominent feature of transgenerational epigenetic inheritance is its non-Mendelian nature: not all offspring that inherit the genetic locus also inherit the parental epigenetic state. Transgenerational epigenetic inheritance is emerging as an important facet of mammalian biology. It may underlie the etiology of human diseases that display complex patterns of inheritance, including diabetes, mental illnesses and autoimmune diseases. As variable epigenetic states can be inherited on an invariant genotype, epigenetic variation may provide a substrate for Darwinian selection that is independent of genetic variation.

X chromosome inactivation is the method of dosage compensation that has evolved to equalise expression of X-linked genes between female (XX) and male (XY) mammals. In somatic cells only one X chromosome is active; the second X in female cells is silenced early during embryonic development, a process that involves the co-ordination of multiple levels of epigenetic regulation to ensure stable chromosome-wide silencing. In this chapter we shall focus on the molecular mechanisms involved in X chromosome inactivation, discuss how the epigenetic marks are believed to elicit stable transcriptional silencing, and why X inactivation represents an excellent model system for studying.

Polycomb-group (PcG) complexes are essential regulators of plant development. These multiprotein complexes repress gene expression by establishing and maintaining trimethylation of lysine 27 at histone H3, a modification that is associated with repressive chromatin. Recent studies have indicated that plant PcG complexes regulate key genes involved in responses to low temperature. Vernalisation is a long-term response to low temperatures whereby plants coordinate their seasonal flowering to occur after winter. In contrast, acclimation of plants to low temperatures, a key step in the establishment of frost tolerance, involves rapid activation of cold-acclimation genes. In this chapter, we describe the dynamics of PcG-mediated gene regulation underlying these two important agronomic traits that are triggered by low temperatures.

In eukaryotes, each chromosome has one centromere and its ends are protected by telomeres. The centromere is a specialised chromosomal locus that directs kinetochore assembly and provides the site for microtubule attachment, allowing accurate chromosome segregation during cell division. Despite the critical role centromeres play, centromeric DNA sequences are highly variable and not conserved between species. Increasing evidence, including the discovery of functional neocentromeres, suggests that centromere identity and function is epigenetically defined through the formation of a specialised chromatin structure.

This chapter reviews recent studies addressing the structural and functional characterisation of centromere chromatin, its assembly and propagation during cell division. Telomeres are specialised nucleoprotein complexes that protect the chromosome ends from degradation. In recent years, it has become increasingly clear that heterochromatic marks at telomeres act as negative epigenetic regulators of telomere elongation, repress recombination events at the telomere and are critical for maintaining telomere structural integrity. Recent research reporting telomeres being transcribed by RNA polymerase II to give rise to TERRA RNA, open up an additional level of regulation at the telomere. This chapter will discuss the links between the epigenetic status of telomeres, telomere function and telomere-length regulation, and the implications on cellular reprogramming, aging and cancer.

DNA in eukaryotes is efficiently and compactly organised into chromatin, the fundamental subunit of which is the nucleosome: approximately 150 bp of DNA spooled 1.65 times around a histone octamer. The location and density of nucleosomes play a role in regulating nuclear processes including transcription, replication, recombination, and repair. Mechanisms acting *in trans*, like ATP-dependent remodelers and cellular memory complexes, as well as *in cis* features intrinsic to the DNA sequence itself regulate the location and density of nucleosomes. Here, we review the three cis acting DNA sequence features that affect the distribution of nucleosomes: (1) two frameworks defining the relationship between the histone octamer and the underlying DNA sequence (nucleosome occupancy and nucleosome position, then statistical positioning and a nucleosome positioning code), (2) the organisation of DNA into the nucleosome core particle, and (3) specific DNA sequence features and DNA templates that promote or inhibit the formation of nucleosomes. We close by describing three computational algorithms trained on DNA sequence that have been used to predict nucleosome position and density. In summary, we hope to draw attention to multiple aspects of DNA sequence that specify organisation of sequence into nucleosomes and influence the distribution of nucleosomes in eukaryotic genomes.

The recent completion of several mammalian genome sequences makes obvious that we share a near-identical collection of genes. What defines us as human must therefore be encoded within regions of the genome where we differ, providing an added level of complexity that probably influences the spatial and temporal expression of genes. Most DNA sequence variation occurs within the repetitive DNA, once called 'Junk DNA' that accounts for at least half of the human genome, and evidence is mounting for its important role in genome function. Although some repeat elements are conserved to some extent between mammals, their precise copy number and genomic location typically are not.

In addition, some repeats are not conserved, including the large tandem repeats. This chapter focuses on two large tandem arrays in the human genome that can adopt quite different chromatin configurations as a result of epigenetic changes; one as a direct consequence of X chromosome inactivation and the other in the context of disease susceptibility. Both cases highlight how alternate packaging of these unusual DNA sequences probably results in differing functions. In each instance, common denominators are the acquisition of the epigenetic organiser protein CTCF and a distinct change in transcripts originating from the array.

In eukaryotes, the genetic material in the form of DNA is wrapped around histone proteins to form nucleoprotein filaments called chromatin. Histones help package the DNA to fit it inside the nucleus of each cell, which in turn regulates access to the genetic information contained within the DNA. Hence, all DNA transactions are likely to be affected by histone metabolism. Eukaryotes carry multiple histones genes that can potentially generate enormous quantities of histone proteins. When present in excess, the positively charged histones can potentially "stick" non-specifically to the negatively charged DNA and adversely affect processes that require access to DNA. Not surprisingly, aberrant histone stoichiometry, chromatin assembly or chromatin structure lead to genomic instability, which is characterised by the increased rate of acquisition of alterations in the genome and is associated with deleterious human conditions such as cancer and aging. Hence, histone synthesis is coupled to ongoing DNA replication and is regulated transcriptionally and post-transcriptionally. To further avoid the deleterious consequences of excess histones, a post-translational regulatory mechanism was described recently whereby excess histones are targeted for degradation by the ubiquitin-proteasome system. In this chapter, we discuss the causes and consequences of excess histone accumulation, as well as the strategies that cells have evolved to deal with them.

Cell-fate specification and stem-cell renewal are fundamental processes in the development of multicellular organisms. In both animals and plants, a key role for transcription factors in these processes has been established, but an accumulating body of evidence indicates that epigenetic regulation also plays a critical role. Once regarded as stable marks, all epigenetic modifications, including DNA and histone methylation, are now known to be reversible, and a cohort of enzymes that add or erase epigenetic marks has been identified. Throughout the life of an organism, the epigenome is dynamically modified, leading in turn to transcriptional changes and ultimately cell-fate specification. Here, I review the recent literature that has shaped this view. Plants are unique among complex organisms in having the ability to generate a whole new organism from a single differentiated cell. How plant cells retain this amazing capability while maintaining their cell identity is a mystery. I introduce a model system for study of cell-fate specification, stem-cell renewal, and reprogramming.

DNA methylation is an important regulator of gene transcription and a large body of evidence has demonstrated that aberrant DNA methylation is associated with unscheduled gene silencing, and the genes with high levels of 5-methylcytosine in their promoter region are transcriptionally silent. DNA methylation is essential during embryonic development, and in somatic cells, patterns of DNA methylation are generally transmitted to daughter cells with a high fidelity. Aberrant DNA methylation patterns have been associated with a large number of human malignancies and found in two distinct forms: hypermethylation and hypomethylation compared to normal tissue. Hypermethylation is one of the major epigenetic modifications that repress transcription via promoter region of tumour suppressor genes. Hypermethylation typically occurs at CpG islands in the promoter region and is associated with gene inactivation. Global hypomethylation has also been implicated in the development and progression of cancer through different mechanisms. This chapter will focus on DNA methylation as the major epigenetic mechanism involved in normal biological processes and abnormal events leading to cancer development. It will also focus on the interaction between DNA methylation and other epigenetic mechanisms.

The histone variants of the H2A family are highly conserved in mammals, playing critical roles in regulating many nuclear processes by altering chromatin structure. One of the key H2A variants, H2A.X, marks DNA damage, facilitating the recruitment of DNA repair proteins to restore genomic integrity. Another variant, H2A.Z, plays an important role in both gene activation and repression. A high level of H2A.Z expression is ubiquitously detected in many cancers and is significantly associated with cellular proliferation and genomic instability. This review summarises the current understanding of these variants and their functions, as well as their links to cancer development. Furthermore, the significance of dysfunction of these variants is highlighted with respect to their potential as biomarkers and as new targets for anticancer therapy.

A wealth of nucleobase and ribose modifications have been identified in multiple types of RNA including tRNAs, rRNAs, mRNAs, and small regulatory RNAs. Among them, 5-methylcytosine (m^5C) has been detected in rRNAs, tRNAs, and early reports have indicated its presence in mRNAs. Well established as an epigenetic mark in DNA, the prevalence and function of m^5C in RNA is either incompletely explored (tRNA, rRNA) or virtually unknown (mRNA, other noncoding RNA). Two eukaryotic m^5C RNA methyltransferases have been identified; however, their substrate specificity and biological roles are incompletely understood. With recent advances in bisulfite sequencing of RNA, comprehensive analyses to determine the occurrence and functions of m^5C in the transcriptome now appear feasible. In this chapter, we summarise the current knowledge in this field, focussing primarily on eukaryotic transcriptomes.

Paramutation is a fascinating phenomenon in which epigenetic information can be transmitted through *trans*-interactions between one allele of a gene to another allele or between homologous DNA sequences that establishes a state of gene expression that is heritable for generations. Paramutation was discovered in maize and similar phenomena have been described in other plants, fungi and animals. In this chapter, we describe several classic plant paramutation systems and discuss recent advances that implicate a role for RNA and a number of components of an RNA-based transcriptional silencing pathway on paramutation.

DNMT3L (DNA methyltransferase 3 like) is member of the DNA methyltransferase family of enzymes responsible for the methylation of CpG dinucleotides. Biochemical studies have revealed that while DNMT3L lacks DNA methyltransferase activity, it can bind to and stimulate the activity of *de novo* DNA methyltransferases DNMT3A and DNMT3B. DNMT3L has also been shown to interact directly with chromatin via its plant homeodomain (PHD) like zinc finger domain. Studies in *Dnmt3L*-deficient mice have revealed that DNMT3L is essential for establishing correct methylation patterns at Retro Transposable Elements (RTE), unique loci and parentally imprinted genes in germ cells, and mice without DNMT3L are rendered infertile. Female *Dnmt3L*$^{-/-}$ mice have apparently normal meiosis but in male *Dnmt3L-/-* germ cells there is asynapsis of chromosomes, and "meiotic catastrophe". Dnmt3L was among the first mammalian genes shown to have a paternal effect, where the genotype of the father (*Dnmt3L*$^{+/-}$) affects sex chromosome aneuploidy in adult and embryonic offspring. This chapter will discuss the role of DNMT3L in establishing DNA methylation patterns during gametogenesis, as well as the proven and potential consequences of DNMT3L deficiency to fertility and somatic and germline genetic disease in light of the increasing evidence that epigenetic reprogramming is a dose sensitive and partially stochastic process.

In the past decade, we have become acquainted with an entire new world of fine regulatory control within cells governed by non-coding RNAs. Although we have not completely explored this world, we know from a few well studied examples that not only gene dosage but protein function also, is fine tuned by non-coding RNAs. It appears that proper cell function is largely dependent on non-coding RNAs, many of which were invisible to us until the advent of high throughput sequencing and tiling array technology. The realisation, that transcripts with no open reading frames are biologically active molecules with multiple functions including regulation of the expression of coding RNA, shifts the power from proteins to non-coding RNAs as key modulators of cellular function. In this review, we discuss the various classes of non coding RNAs and the mechanisms employed by them to achieve this status, with a particular focus on their ability to induce epigenetic modifications.

A large variety of methods to measure DNA methylation have been developed and used extensively over the last twenty years. These have been based on selective restriction digestion of methylated DNA, the capture of methylated DNA by methyl binding proteins or antibodies, or bisulphite conversion of DNA. However, all restriction enzyme based methods are dependent on available restriction sites for methylation-specific restriction enzymes and therefore cannot be used to analyse every CpG site in the genome. Although immunoprecipitation methods are not sequence specific, they are unable to provide methylation data at a single-base resolution. Therefore, both of these approaches are limited in their applicability. On the other hand, bisulphite conversion based methods allow methylation studies directed to any CpG site in the genome.

Sodium bisulphite treatment of DNA converts all unmethylated cytosines into uracil while methylated cytosines remain unchanged, thus transferring an epigenetic difference into a measurable genetic difference. A variety of downstream methods such as Polymerase Chain Reaction (PCR), sequencing, Single Nucleotide Polymorphism (SNP) genotyping and mass spectrometry can be coupled with bisulphite conversion. This chapter provides an overview of some of these methods, concentrating on bisulphite sequencing, methylation-specific PCR, pyrosequencing, MassARRAY EpiTYPER and Infinium Human Methylation27 Bead Chip. Considerations for assay selection and detailed protocols for each method are presented in the end of this chapter.

Methylation-sensitive high resolution melting (MS-HRM) is an inexpensive and robust closed tube screening methodology that enables rapid analysis of locus specific DNA methylation for multiple samples. MS-HRM is based on the differential melting behaviour of PCR amplification products derived from methylated and unmethylated templates after bisulfite treatment. The melting profiles of an unknown sample are compared to the melting profiles of standards with known DNA methylation levels. MS-HRM has the advantage of allowing ready distinction between homogenous and heterogeneous DNA methylation. Estimation of DNA methylation to quite low levels in a semi-quantitative manner is possible for homogeneously methylated templates where all the CpG sites are methylated. However for heterogeneous methylation, the formation of multiple heteroduplexes makes quantitation difficult. Digital MS-HRM utilises limiting dilution to amplify single templates enabling DNA methylation analysis at a single allele resolution and enables the quantitative analysis of both homogenous and heterogeneous methylation. The PCR products either generated by MS-HRM or digital MS-HRM can subsequently undergo Sanger DNA sequencing or pyrosequencing to investigate DNA methylation at individual CpG positions.

The location and density of nucleosomes in the eukaryotic genome plays a role in regulating nuclear processes including transcription, replication, recombination, and repair. Microarray mapping of nucleosomally protected DNA has emerged as a powerful, cost-effective, high-throughput method to analyse the relationship between nucleosome position and genomic regulation. In this chapter we discuss experimental considerations such as sample preparation and microarray design. In addition, two procedures are detailed: (1) Formaldehyde Crosslink and Harvest Cells, Isolate Nuclei, and MNase cut chromatin, and (2) Isolation of mononucleosomally protected DNA and fluorescent labelling of DNA for microarray hybridisation. With the information specified in this chapter, most any laboratory equipped for molecular biology with institutional or commercial access to microarray facilities should be should be able to map nucleosome position and occupancy.

DNA methylation plays an essential role in normal human development, where abnormalities in proper establishment and maintenance of DNA methylation patterns result in human disease. Many experimental approaches have been developed for assaying DNA methylation patterns, including enzymatic-based approaches. In this chapter, we highlight some of these approaches and describe their relative advantages and disadvantages. We also describe advances in microarray and sequencing technologies that have improved resolution of enzymatic-based methods, providing expanded coverage of CpG dinucleotides throughout the genome. These approaches are important tools in characterising the role of DNA methylation in genome organisation and function.

With the advances in traditional genetics failing to provide causal genes for many complex diseases, the focus of research is shifting towards determining the importance of gene-environment interactions, more specifically epigenetic regulation. Paramount to answering this question is the knowledge of which transcription factors bind to what sequence, as well as a detailed understanding of how the transcriptional state of a genetic sequence is epigenetically distinguished. The chromatin immunoprecipitation (ChIP) strategy has proven to be a powerful tool to investigate these mechanisms, but has been limited for a long time to single locus analysis. The recent emergence of next-generation sequencing (NGS) technology however has revolutionised the field of epigenomics and for the first time enabled unbiased genome-wide analysis of ChIPed DNA. Here we provide a detailed discussion and protocol on how to best perform ChIP for NGS analysis.

Over the past decades it has become ever more apparent that understanding the genome did not stop with unravelling the genetic code. Regulatory mechanisms are needed to determine which parts of the genome are active or inactive, and to form a memory system that can be passed on over multiple cell divisions for proper functioning of the cell. These mechanisms underpinning heritable gene regulation are encapsulated by the term "epigenetics" and include histone modifications, miRNA and non-coding RNA expression, and methylation of cytosine residues in DNA. In this review, we compare the various methods that can be used to analyse DNA methylation patterns throughout the genome, and discuss their advantages and disadvantages. In addition, we present a detailed protocol for genome-wide DNA methylation analysis based on the capture of methylated DNA using a methyl-CpG binding domain-based (MBD) protein combined with second generation sequencing.

Since biological phenotype and differentiation is regulated partially through CpG DNA methylation, and this mark is relatively easy to measure, genome-wide profiling of methylation landscape is a popular tool in epigenetic research. The burden then falls on bioinformatics to provide normalisation, quality control, and interpretation, while adopting to the vast number of versatile methods to interrogate methylation profiles. While creating a relational report of the results from either of these methods is critical for data to cross over from lab to lab, and from research to diagnostic and translation purposes, the methods are each introducing their unique bias to the data.

Ideally, one would hope all labs would adopt a single method, but since each method offers a unique advantage, such as price, sensitivity, or confidence, one has to overcome the disparity hurdle via bioinformatic tools. Thus identifying the methodological biases of each technique, and the way to compensate for those in the process of generating a universal CpG methylation prediction on a genome region is the ultimate goal we are describing here. Follow up of CpG methylation with other read outs, such as impact on gene expression or coincidence with locations in the genome, where allelic variation is associated with the investigated phenotype, are also key to proper interpretation of the results.

Maize has served as an excellent model for the study of epigenetic gene regulation for the past several decades. The pioneering work of maize geneticists like Barbara McClintock, Alexander Brink, Marcus Rhoades, and others led to the observation of many fascinating phenomena that were later demonstrated to be epigenetically regulated events. Since these observations were made, a great deal of progress has been made in determining the underlying causes of the phenomena, and many of these examples of epigenetic gene regulation are currently being used to elucidate the mechanisms of epigenetic heritability in plants. Today, these phenomena and the mutants that impact them serve as resources for studying how DNA methylation, chromatin structure, and small RNAs act to influence paramutation, gene silencing, and parent-of-origin dependent dosage compensation. The key attributes and potential contributions of each resource are discussed in the context of understanding the mechanisms and significance of epigenetic gene regulation in large, complex genomes.

Biological experiments are rapidly moving into the information age with the explosion of data generated from rapidly evolving technologies. Microarrays can interrogate many thousands to millions of loci within any one sample, while massively parallel sequencing platforms can essentially measure the entire genome. Keeping up with this rapid pace of data accumulation has required the development of online tools for the processing, analysis and annotation of data generated from large numbers of microarrays or next generation sequencing (NGS) experiments. I will cover a selected range of tools available to the biological researcher, from online discussion forums and blogs, to curated databases that store the data and associated annotations. I will then cover viewer tools that enable visualisation the annotations and finally review tools for assay design for follow up validation of NGS and microarray analyses. I will pay particular attention to DNA methylation and provide the reader with an insight into what is available online for the epigenetics researcher aiming to make sense of epigenomic data.

Epigenetics can appear as an impenetrable subject; not just to those encountering it for the first time, but to those within the field too. However, epigenetics, like any subject can be made easier to understand using a combination of clear language, creative illustrations and even animations and film clips. This chapter aims to point readers of all experiences towards helpful and easy-to-read resources that educate about epigenetics. It is split into two main sections, the first aimed at a lay audience including teachers and high school students and the second, at graduate and postgraduate students and beyond. Each section contains summaries of published articles and web sites. The chapter ends with a short section on epigenetic societies and research networks and a summary table of resources.

It is intended to provide a sample of some of the best short to medium length reviews on general topics within the field of epigenetics and while we cover a wide variety of themes, we apologise for any areas not covered. We cite the URLs of freely-available articles wherever possible, but many articles will require library access. We also urge readers to contact authors or publishers if they wish to distribute any of the articles for teaching purposes.

Systems Microbiology

Systems biology is the study of the dynamic interactions of more than one component in a biological system in order to understand and predict the behaviour of the system as a whole. Systems biology is a rapidly expanding discipline fuelled by the 'omics era and new technological advances that have increased the precision of data obtainable. A focus on simple single cell organisms such as bacteria aids tractability and means that systems microbiology is a rapidly maturing science.

This volume contains cutting-edge reviews by world-leading experts on the systems biology of microorganisms. As well as covering theoretical approaches and mathematical modelling this book includes case studies on single microbial species of bacteria and archaea, and explores the systems analysis of microbial phenomena such as chemotaxis and phagocytosis.

Topics covered include mathematical models for systems biology, systems biology of *Escherichia coli* metabolism, bacterial chemotaxis, systems biology of infection, host-microbe interactions, phagocytosis, system-level study of metabolism in *Mycobacterium tuberculosis*, and the systems biology of *Sulfolobus*.

This book is a major resource for anyone interested in systems biology and a recommended text for all microbiology laboratories.

Modelling methodologies in the life- and biomedical sciences are hampered by the complexity of the processes and systems at work. Modelling studies into prokaryotic systems require the elucidation of the mechanistic model. In this chapter we introduction modelling methodologies and discuss the problem of model (and parameter) inference. We comment on state-of-the-art research questions and provide a general discussion on how models can and should be used in order to better understand the structure, function and dynamics of biological systems. The aim is not to provide an introduction to modelling per se, but to provide readers with an overview on the available methodologies. The modelling approach chosen depends on the biological question at hand as well as a range of social factors.

The functional repertoire of an organism's metabolic network is closely linked to its phenotype and potential for utility in metabolic engineering applications. In this chapter, we discuss a systems biology view of *Escherichia coli* metabolism by integrating current genome-scale computational modelling approaches with available molecular genetics tools, as well as the experimental framework for metabolite and metabolic flux determination.

Bacterial chemotaxis is a paradigm for biological sensing and information transmission. The chemotaxis signal-transduction pathway allows cells to sense chemicals in their surroundings in order to regulate flagellated rotary motors, thus allowing them to swim towards nutrients and away from toxins. Importantly, cells are able to sense with remarkably high sensitivity over a wide range of chemical background concentrations. To make this possible, chemoreceptors do not signal independently but form clusters for amplification and integration of signals, as well as for adaptation to persistent stimulation.

While chemotaxis in *Escherichia coli* has been exceptionally well characterised, new experimental facts still require revisions of existing models and thus further increase our understanding of sensing and signalling in bacteria. Additionally, experiments on other bacterial species such as *Bacillus subtilis* and *Rhodobacter sphaeroides* indicate that bacteria other than *E. coli* can have substantially different and more complex chemotaxis pathways, which provides renewed challenges for experimentalists and modellers alike. Here we discuss our current understanding as well as the frontiers of bacterial chemotaxis research.

Microbial infections still cause around one quarter of all deaths globally, despite the advances that have been made in the treatment of infectious disease. The increasing occurrence of drug resistant pathogens, both old and new, coupled with an increasingly mobile human population has created many novel opportunities for potential pathogens to meet new human hosts. All of this requires new prevention and control strategies but progress has been slow, despite recent technological advances and increased investment. The rapid increase in data proved difficult to translate into practical applications for human health care, and new approaches and analyses are needed to make the most of new opportunities. One of these is the use of the new tools of systems biology and this chapter will review the application of these to microbial pathogens.

Host-microbe interactions are complex phenomena spanning multiple levels of complexity, from environmental and ecological factors up to the cellular and genetic levels of host responses. At each of these levels a relationship is established between one or more microorganisms and the host, resulting in formation of various forms of associations ranging from symbiosis to parasitism. Pathogens have the potential to cause disease in their hosts through host-pathogen interactions in which host defences are challenged by the invasive capacities of the pathogen. The aim of this chapter is to give an outline of attempts made to unravel the components of host-pathogen interactions at the cellular and molecular levels and to discuss strategies to skew the balance in favour of the host, focusing our attention on crucial intracellular pathogens causing globally relevant diseases such as tuberculosis, gastroenteritis, influenza and malaria.

Phagocytosis is the fundamental cellular process by which eukaryotic cells bind and engulf particles by deforming their plasma membrane. Particle engulfment involves particle recognition by cell-surface receptors, signalling, and remodelling of the actin cytoskeleton to guide the membrane around the particle in a zipper-like fashion. The signalling complexity is daunting, involving hundreds of different molecular species during the initial stages of engulfment. For instance, the well-characterised immune Fcã and the integrin CR3 receptors signal to tyrosine kinases and Rho GTPases, ultimately regulating a wide variety of proteins, which direct actin polymerisation and myosin-motor proteins for force generation and contraction. Despite the signalling complexity, phagocytosis also depends strongly on simple biophysical parameters, such as shape and cell stiffness, or membrane biophysical properties that are independent of the type of cell or particle. We argue that these emergent, universal features are particularly important to address in order to explain this evolutionary well-conserved and robust mechanism. In this chapter we review these universal features to describe the principles of phagocytosis. Specifically, we use a recently published model of phagocytic engulfment as a guide. Finally, we discuss state-of-the-art live-cell fluorescence microscopy, recently used to elucidate the dynamics of phospholipids, actin polymerisation and myosins in the particle-shape recognition by the amoeba *Dictyostelium*.

Despite decades of research many aspects of the biology of *Mycobacterium tuberculosis* remain unclear and this is reflected in the antiquated tools available to treat and prevent tuberculosis. Consequently, this disease remains a serious public health problem responsible for 2 to 3 million deaths each year. Important discoveries linking *M. tuberculosis* metabolism and pathogenesis have renewed interest in the metabolic underpinning of the interaction between the pathogen and its host. Whereas, previous experimental studies tended to focus on the role of single genes, antigens or enzymes the central paradigm of systems biology is that the role of any gene cannot be determined in isolation from its context.

Therefore systems approaches examine the role of genes and proteins embedded within a network of interactions. We here examine the application of this approach to studying metabolism of *M. tuberculosis*. Recent advances in high throughput experimental technologies, such as functional genomics and metabolomics, provide datasets that can be analysed with computational tools such as flux balance analysis. These new approaches allow metabolism to be studied on a genome scale and have already been applied to gain insights into the metabolic pathways utilised by *M. tuberculosis in vitro* and identify potential drug targets. The information from these studies will fundamentally change our approach to tuberculosis research and lead to new targets for therapeutic drugs and vaccines.

Life at high temperature challenges the stability of macromolecules and cellular components, but also the stability of metabolites, which has received little attention. For the cell, the thermal instability of metabolites means it has to deal with the loss of free energy and carbon, or in more extremes, it might result in the accumulation of dead-end compounds. In order to elucidate the requirements and principles of metabolism at high temperature, we used a comparative blueprint modelling approach of the lower part of the glycolysis cycle. The conversion of glyceraldehyde 3-phosphate to pyruvate from the thermoacidophilic Crenarchaeon *Sulfolobus solfataricus* P2 (optimal growth-temperature 80°C) was modelled based on the available blueprint model of the eukaryotic model organism *Saccharomyces cerevisiae* (optimal growth-temperature of 30°C).

In *S. solfataricus* only one reaction is different, namely glyceraldehyde-3-phosphate is directly converted into 3-phosphoglycerate by the non-phosphorylating glyceraldehyde-3-phosphate dehydrogenase, omitting the extremely heat-instable 1,3-bisphosphoglycerate. By taking the temperature dependent non-enzymatic (spontaneous) degradation of 1,3-bisphosphoglycerate in account, modelling reveals that a hot lifestyle requires a cool design.

Chapter 3

Essentials of Veterinary Parasitology

Approaches to the teaching of veterinary parasitology face two major challenges. First, the quantity of data describing any given parasite can be overwhelming, if not indigestible, for students. Second is the urge to write more and more about less and less, which is the bane of those who write textbooks intended to be used by students. To meet these challenges the editors of this volume have opted to be selective in the choice of topics in an effort to make the book readable, rather than comprehensive.

Essentials of Veterinary Parasitology provides an up-to-date resource for students and practicing veterinarians on how to recognise, diagnose and treat parasitic diseases in livestock and companion animals. Featuring full-colour illustrations and a user friendly layout, it begins with a section dedicated to the fundamentals of veterinary parasitology and ends with a section on the prevention of parasitic infections entailing recent developments in our understanding of the pathogenesis and control of parasitic diseases. In-between are sections on important parasitic infections in livestock organised by the parasite agents - helminths, protozoa and arthropods - plus a section on diagnostic parasitology. This book is an essential reference for veterinary students, practicing veterinarians and researchers in the field of parasitology.

We tend to think of parasites as a nuisance, but they are in fact very serious disease-causing agents. Despite advances of veterinary medicine, parasitic diseases have remained a major cause of morbidity, mortality and economic losses, worldwide. With the increasing burden of parasites on human and animal suffering, study of "parasitology" has become an important and rapidly growing discipline of science. Veterinarians' awareness of parasitic diseases is undoubtedly more critical now than at any time in the history of veterinary medical practice. This chapter provides a short introduction to parasites and their unique properties.

How do parasites establish themselves inside their hosts is one of the most intriguing questions in parasitology. Given the diversity that exists among parasites, it is not surprising that the mechanics of parasitic infections are highly variable. The diverse strategies that pathogenic parasites use to infect their hosts have become better understood by means of molecular techniques that have allowed the identification of parasitic genes and virulence factors which are crucial to disease establishment and progression. Despite all this progress, predicting and derailing the transmission of any parasitic disease remains a challenge. This chapter addresses the interplay between parasites and their host organisms, introduces the fundamentals of parasite infection, and provides an overview of the multiple determinants - related to the parasite, the host, and the environment - that influence the parasite's capacity to cause a disease at the cellular and population levels.

In addition to the host's behaviour and physiology, immunity is another important host factor that influences the host-parasite interactions. The immune system is one of the most complex and diverse body components. This system recognises parasite antigens as 'non-self' (foreign) and an immune response to these parasites is then initiated. It is difficult to generalise about the mechanisms of anti-parasite immunity because there are many different parasites that have different forms, which reside in different tissue of the animal body and in some cases may reside in different host species. Parasites have also evolved very elaborate mechanisms to escape attack by the host immune system. In this chapter we describe the range of defences the mammalian body possess to combat parasitic infections.

Classification of nematodes has been traditionally based on the presence or absence of a posterior cuticular chemoreceptor called 'phasmid'. Nematode species with phasmid are known as phasmidea (Secernentea) and nematodes that lack phasmid are called aphasmidea (Adenophera). It is important to realise that the parasite taxonomy is an evolving field and there is no a single scheme that is always acceptable. Class Nematoda encompasses numerous species that infect livestock and companion animals. This chapter focuses only on the most economically important nematode infections in livestock and companion animals. General taxonomy of nematodes considered in this chapter is given to the genus level.

Cestoda is a class of parasitic flatworms (Platyhelminthes), commonly called tapeworms or cestodes. All tapeworms use vertebrates as a definitive host, and vertebrates or invertebrates (arthropods, crustaceans) as an intermediate host, depending on the species. The definitive host harbours the adult, sexual, or mature stages of parasite. Larval 'metacestode' development occurs in the intermediate host (I.H.), which will be eaten by definitive host. In the latter, larval stages attach to the gut mucosa and mature to adult tapeworms via a process called 'strobilation'. Most tapeworms are found in the small intestine of their host as adults or, with *Thysanosoma* spp., have access to the intestine.

Dicrocoeliosis is caused by *Dicrocoelium dendriticum*, which is also known as 'lancet fluke' or 'small liver fluke'. It can infect sheep, goats, cattle, deer and rabbits, and occasionally horses and pigs. Dicrocoeliasis is a widespread problem worldwide in grazing livestock. The epidemiology of Dicrocoelium depends upon the environment and on the presence of its intermediate and definitive hosts. *Dicrocoelium* spp. do pose a zoonotic risk but are very uncommon in humans, most cases are likely to be non-symptomatic.

This chapter discusses *Acanthamoeba* granulomatous encephalitis, amoebiasis, babesiosis, *Balamuthia* amoebic encephalitis, balantidiasis (also known as balantidiosis), besnoitiosis, blastocystosis, coccidiosis, cryptosporidiosis, cyclosporiasis, cytauxzoonosis, equine protozoal myeloencephalitis, giardiosis, leishmaniasis, neosporiosis, primary amoebic meningoencephalitis, rhinosporidiosis, sarcosporidiosis, theilerioses, mediterranean coast fever, trichomoniosis, toxoplasmosis, trypanosomiosis, nagana or african trypanosomiosis.

Mites are ectoparasites of a wide range of birds, domesticated and wild animal species. Some have a zoonotic significance. Mites are members of the phylum Arthropoda. *Demodex* spp and *Psorobia* spp are host specific, and these species will not cross-infest other hosts. However mange mites (*Chorioptes* spp, *Psoroptes* spp and *Sarcoptes* spp) are no host specific and can cross-infest a large number of hosts. Mites live on the host continuously and infest other animals by contact. The life cycles of mites are all slightly different because some burrow, whereas others live on the surface of the skin. *Sarcoptes* spp. and *Notoedres cati* females burrow in the skin and deposit eggs. The eggs hatch into six-legged larvae, which develop and molt to eight-legged protonymphs and tritonymphs, which develop and molt into adults. The entire cycle requires 9 to 17 days. The ticks found on domesticated animals are not host specific, although they do have host preferences, and their distribution is subject to environmental conditions. Ticks are identified as being soft or hard ticks. The hard ticks are generally classified as one-, two-, or three-host ticks. Three-host ticks may complete the cycle in a short period (*Rhipicephalus* spp.), whereas other ticks (*Dermacentor* spp.) require 2 years, with 1 year between each stage before they reattach to a host.

Ticks are giant acarids (phylum Arthropoda), which have a major veterinary and public health impact. They represent an obstacle in economic growth especially in developing countries. Due to their feeding behaviour ticks inflict considerable physical damage and irritation which disrupt the foraging of livestock, thereby reducing productivity and fitness, and lowering defences against other diseases (e.g. tick-borne fever predisposes lambs to tick pyaemia). Wounds induced by tick bites are open to invasion by secondary bacterial and fungal, and other opportunistic infections. Tick infestation may also cause tick paralysis, thought to be due to a neurotoxin elaborated by the tick's ovaries, and introduced into the host with saliva while the tick is feeding. This condition is generally characterised by progressive, ascending, flaccid motor paralysis with muscle incoordination and ataxia.

Accurate diagnosis of parasitic infections is a prerequisite for successful treatment and control of these pathogens. Errors in the diagnosis can lead to the initiation of unnecessary therapies, or delays in initiating the correct therapy. Thus, the clinicians must maintain a sharp index of suspicion and must rely on detailed history and clinical manifestations, to raise the possibility of a parasitic disease. Even though the diagnosis can be difficult, and definitive identification of the parasites can be challenging particularly in the non-endemic settings. Therefore, laboratory testing for detection and identification of the parasitic agents is required to complement clinical judgement, enhance the clinician's ability to select specific antiparasitic drugs, and ultimately improve patient care. A wide range of laboratory procedures are available for the diagnosis of parasitic infections. These procedures vary in methodology, expense, availability, sensitivity, and specificity. In this chapter, the standard techniques used in the laboratory diagnosis of parasitic infections are discussed.

Parasites may induce a wide variety of pathology in their host tissues. These changes vary from inapparent, to frank necrosis, grossly visible granulomas, and induction of hyperplastic or neoplastic changes in various tissues. The host may exhibit no clinical signs, or develop anemia, hypoproteinemia, weight loss, anorexia, even death. A few simple terms need to be defined before we begin our discussion of pathology. Localised infection refers to a parasitic infection that is limited to a single host tissue or focal areas of a given organ system. While generalised parasitic infection refers to parasites that have widespread migration throughout the host body in many tissues. While we are not going to make the reader a pathology expert, you should understand some basic pathology terms. Necrosis refers to destruction of normal tissue cells and organ architecture resulting in accumulations of cellular debris, fibrin, inflammatory cells, and red blood cells.

Granulomatous inflammation is a chronic form of inflammation which is comprised predominantly of mixed mononuclear leukocytes including macrophages, lymphocytes, plasma cells, and sometimes multinucleated giant cells. If a granuloma has central zone of necrosis, this is a caseogranuloma. A hyperplastic change involved increased numbers or size of normal tissues, such as thickening of a keratinised layer in the epidermis being known as hyperkeratosis. Neoplastic transformation means that a true neoplasm consisting of a monomorphic population of cells has developed into a microscopic or grossly visible tumour or mass.

Most animals will carry a few parasites in/on their body and this is normal if kept under control. But, if the infection becomes overwhelming the animal's health will suffer and irreversible damage could result. Indeed, parasite infections continue to be one of the most economically important constraints in raising livestock worldwide, a significant health and welfare issue in companion animals, and an important source of zoonotic infections in humans. Despite tremendous efforts, the number of eradicated parasites is negligible and the perspectives for future eradications would most likely be counteracted by the emergence or re-emergence of other parasite species.

Control of parasites can be challenging because parasites can use different immune evasive strategies and/or become resistant to drugs following exposure to the host immune response or to non-judicious use of antiparasitic therapy, respectively. These challenges necessitate an integrated parasite management approach that encompasses a range of manipulations of the host (e.g. increased genetic host resistance through selecting for low fecal worm egg count, improved host resistance through proper nutrition), the environment (e.g. pasture management, appropriate husbandry, sound sanitation) and the parasite (e.g. sensible use of antiparasitics, maintain susceptible population of parasites, sterile male technologies for insects). This chapter will consider the reasons for parasite control followed by general considerations for parasite control and finally specific considerations for control of endoparasites and ectoparasites in ruminants, horses and companion animals.

Parasites such as nematodes and mites can be debilitating and deadly inhabitants of an animal's body. While some parasitic infections can be controlled effectively by preventive biosecurity, vaccines or other non-pharmaceutical intervention measures, for many parasites, these measures are not available, have a limited effect, or cannot be applied in practical settings. Antiparasitic drugs are the commonly-applied pharmaceutical compounds used to reduce, treat or prevent parasitic infections in animals.

Starting perhaps with the initial usage of carbon tetrachloride against Fasciola in cattle, the strategic use of anthelmintic drugs has drastically reduced gastro-intestinal helminth infections and improved the welfare and productivity of domestic animals. Widespread effective usage of antiparasitics is therefore one of the greatest triumphs of the parasitology discipline, although they are not without problems and shortcomings. One of the most pressing concerns addressed in this chapter is the emergence of strains of parasites that are resistant to the action of an antiparasitic drug.

Anthelmintic resistance is defined as a heritable genetic change in a population of worms that produces an alteration in the chemical sensitivity of that population. This change enables some individual worms in that population to survive drug treatments that are generally effective against the same species and stage of infection at the same dose rate. In practical terms anthelmintic resistance is present in a population of parasites when the efficacy of the drug falls below that which is historically expected, when other causes of reduced efficacy have been ruled out. These types of genetic changes occur slowly, usually over many years, and are the direct result of natural selection on parasite populations in response to drug treatments. Precise molecular mechanisms have not been clearly elucidated for any of the anthelmintics, but some important mutations have been identified that are involved in resistance to benzimidazole drugs.

Synthetic Biology Advance: Genome Transferred between two Bacteria via Yeast

Microbes can be resistant to genetic engineering. There's simply not enough DNA in some of them to permit significant alteration. But by building a bacterial genome inside yeast—a more complex and information-rich eukaryote that is one of mankind's oldest genetic engineering projects—scientists have successfully created new, synthetic bacterial strains, according to a paper published today in Science.

Carole Lartigue and colleagues at the J. Craig Venter Institute have been seeking to build living cells from scratch. The ultimate goal is to produce man-made microbes to solve man-made problems, whether eating up carbon dioxide or making the fuels of the future.

Working with microbes Mycoplasma mycoides and Mycoplasma capricolum (goat pathogens with extremely small genetic blueprints), the scientists previously swapped one genome for the other. And colleagues at JCVI chemically synthesized the genome of Mycoplasma genitalium inside yeast.

Now Lartigue and colleagues report success in synthesizing, cloning and transplanting a M. mycoides genome into M. capricolum cells. The key? Yeast again.

By working inside the roomier (and better scientifically understood) confines of yeast, the researchers can more easily manipulate the synthetic genetic material. This enabled them to clone the Mycoplasma genome inside the yeast one and to build in protections for this synthetic genome from inherent mechanisms that protect cells from foreign DNA—akin to the human immune system rejecting a transplanted organ.

Once transplanted back from the yeast, specially tagged M. mycoides *genomes began to appear in the subsequent cells divided from the host* M. capricolum, *proving they had been accepted by the host. This opens the door to broad tinkering with genomes in yeast by providing a method to get them back into host cells—one step closer to a fully synthetic life form.*

A Circuit in Every Cell

Researchers in nanomedicine have long dreamed of an age when molecular-scale computing devices could be embedded in our bodies to monitor health and treat diseases before they progress. The advantage of such computers, which would be made of biological materials, would lie in their ability to speak the biochemical language of life.

Several research groups have recently reported progress in this field. A team at the California Institute of Technology, writing in the journal Science, made use of DNA nanostructures called seesaw gates to construct logic circuits analoguous to those used in microprocessors. Just as silicon-based components use electric current to represent 1's and 0's, biobased circuits use concentrations of DNA molecules in a test tube. When new DNA strands are added to the test tube as "input," the solution undergoes a cascade of chemical interactions to release different DNA strands as "output." In theory, the input could be a molecular indicator of a disease, and the output could be an appropriate therapeutic molecule. A common problem in constructing a computer in a test tube is that it is hard to control which interactions among molecules occur. The brilliance of the seesaw gate is that a particular gate responds only to particular input DNA strands.

In a subsequent Nature paper, the Caltech researchers showed off the power of their technique by building a DNA-based circuit that could play a simple memory game. A circuit with memory could, if integrated into living cells, recognise and treat complex diseases based on a series of biological clues.

This circuitry has not been integrated into living tissue, however, in part because its ability to communicate with cells is limited. Zhen Xie of the Massachusetts Institute of Technology and his collaborators have recently made progress on this front. As they reported in Science, they designed an RNA-based circuit that was simpler but could still distinguish modified cancer cells from noncancerous cells and, more important, trigger the cancer cells to self-destruct.

Both techniques have been used only in artificial scenarios. Yet the advances in DNA-based circuits offer a new, powerful platform to potentially realise researchers' long-held biocomputing dreams.

Polyester bees are all over the Northeast. The interesting thing about them is that they dig underground tunnels, about the width of your pinky finger, where they lay their eggs. To protect their larvae from heat, cold, fungus, bacteria and other dangers, the bees line these chambers with a clear, cellophanelike substance. The larvae then live underground for most of their lives in these reinforced cells.

I kind of stumbled on polyester, or Colletes, bees somewhere on the Internet and eventually got samples of their nest cells from the American Museum of Natural History in New York City. We haven't published our work yet, but we have been looking at these cells and trying to figure out what they're made of.

The bad news is that these cells are really hard to study because their job is to be hard to break down. We found ourselves in this catch-

22: anything nasty enough to break them down was too nasty to put into our equipment and anything we could put into our equipment wouldn't break them down.

But what we did show was that it's not just plastic. There are actually silk fibres that the bees lay down first, and the plastic is put down on top of the fibres like fibreglass and that makes it really durable. We're working with bacteriologists to see if we can find a bacterium that will break down the plastic.

We care about this material for two reasons. The official reason is that it's a fascinating biologically derived material that isn't biodegradable. So I don't know if you ever do this, but I occasionally forget about spaghetti in the back of my fridge. You wind up with a sealed plastic container that has decomposing stuff inside. You don't want your container to also be part of the decomposing stuff, but we also don't want to keep filling our landfills with containers. This could be a material that's robust under normal circumstances but can be broken down and reused. The second reason I care about this is that it's emblematic of the fact that there's an enormous amount we don't know about the world around us. It makes me wonder how many other things there are like this.

Biological scientists study living organisms and their relationship to their environment. They research problems dealing with life processes and living organisms. Most specialise in some area of biology, such as zoology (the study of animals) or microbiology (the study of microscopic organisms). (Medical scientists' work is closely related to that of biological scientists.)

Many biological scientists work in research and development. Some conduct basic research to advance knowledge of living organisms, including viruses, bacteria, and other infectious agents. Basic biological research continues to provide the building blocks necessary to develop solutions to human health problems and to preserve and repair the natural environment. Biological scientists mostly work independently in private industry, university, or government laboratories, often exploring new areas of research or expanding on specialised research started in graduate school. Those who are not wage and salary workers in private industry typically submit grant proposals to obtain funding for their projects.

Colleges and universities, private industry, and Federal Government agencies such as the National Institutes of Health and the National Science Foundation contribute to the support of scientists whose research proposals are determined to be financially feasible and to have the potential to advance new ideas or processes.

Biological scientists who work in applied research or product development use knowledge provided by basic research to develop new drugs, treatments, and medical diagnostic tests; increase crop yields; and protect and clean up the environment by developing new biofuels. They usually have less autonomy than basic researchers to choose the emphasis of their research, relying instead on market-driven directions based on their firms products and goals. Because biological scientists doing applied research and product development in private industry may be required to describe their research plans or results to nonscientists who are in a position to veto or approve their ideas, they must understand the potential cost of their work and its impact on business. Scientists often work in teams, interacting with engineers, scientists of other disciplines, business managers, and technicians. Some biological scientists also work with customers or suppliers and manage budgets.

Those who conduct research usually work in laboratories and use electron microscopes, computers, thermal cyclers, and a wide variety of other equipment. Some conduct experiments using laboratory animals or greenhouse plants. This is particularly true of botanists, physiologists, and zoologists. For some biological scientists, research also is performed outside of laboratories. For example, a botanist might do research in tropical rain forests to see what plants grow there, or an ecologist might study how a forest area recovers after a fire. Some marine biologists also work outdoors, often on research vessels from which they study various marine organisms such as marine plankton or fish.

Some biological scientists work in managerial or administrative positions, usually after spending some time doing research and learning about a particular firm, agency, or project. They may plan and administer programs for testing foods and drugs, for example, or direct activities at zoos or botanical gardens. Some work as consultants to businesses or to government agencies.

Recent advances in biotechnology and information technology are transforming the industries in which biological scientists work. In the 1980s, swift advances in basic biological knowledge related to genetics and molecules spurred growth in the field of biotechnology. Biological scientists using this technology manipulate the genetic material of animals or plants, attempting to make organisms more productive or resistant to disease. Research using biotechnology techniques, such as recombining DNA, has led to the production of important substances, including human insulin and growth hormone. Many other substances not previously available in large quantities are starting to be produced by biotechnological means; some may be useful in treating cancer and other diseases. Today, many biological scientists are involved in biotechnology. Those who work on the Human Genome Project isolate genes and determine their function.

This work continues to lead to the discovery of the genes associated with specific diseases and inherited traits, such as certain types of cancer or obesity. These advances in biotechnology have created research opportunities in almost all areas of biology, with commercial applications in the food industry, agriculture, and environmental remediation, and in other emerging areas such as DNA fingerprinting.

Most biological scientists are further classified by the type of organism they study or by the specific activity they perform, although recent advances in the understanding of basic life processes at the molecular and cellular levels have blurred some traditional classifications.

Aquatic biologists study micro organisms, plants, and animals living in water. Marine biologists study salt water organisms, and limnologists study fresh water organisms. Much of the work of marine biology centres on molecular biology, the study of the biochemical processes that take place inside living cells. Marine biologists sometimes are mistakenly called oceanographers, but oceanography is the study of the physical characteristics of oceans and the ocean floor.

Biochemists study the chemical composition of living things. They analyse the complex chemical combinations and reactions involved in metabolism, reproduction, growth, and heredity. Biochemists and molecular biologists do most of their work in biotechnology, which involves understanding the complex chemistry of life.

Botanists study plants and their environment. Some study all aspects of plant life, including algae, fungi, lichens, mosses, ferns, conifers, and flowering plants; others specialise in areas such as identification and classification of plants, the structure and function of plant parts, the biochemistry of plant processes, the causes and cures of plant diseases, the interaction of plants with other organisms and the environment, and the geological record of plants.

Microbiologists investigate the growth and characteristics of microscopic organisms such as bacteria, algae, or fungi. Most microbiologists specialise in environmental, food, agricultural, or industrial microbiology; virology (the study of viruses); immunology (the study of mechanisms that fight infections); or bioinformatics (the process of integrating molecular biology and information science). Many microbiologists use biotechnology to advance knowledge of cell reproduction and human disease.

Physiologists study life functions of plants and animals, both in the whole organism and at the cellular or molecular level, under normal and abnormal conditions. Physiologists often specialise in functions such as growth, reproduction, photosynthesis, respiration, or movement, or in the physiology of a certain area or system of the organism.

Biophysicists study the application of principles of physics, such as electrical and mechanical energy and related phenomena, to living cells and organisms.

Zoologists and wildlife biologists study animals and wildlife their origin, behaviour, diseases, and life processes. Some experiment with live animals in controlled or natural surroundings, while others dissect dead animals to study their structure. Zoologists and wildlife biologists also may collect and biological data to determine the environmental effects of current and potential use of land and water areas. Zoologists usually are identified by the animal group studied ornithologists (birds), mammalogists (mammals), herpetologists (reptiles), and ichthyologists (fish).

Ecologists study the relationships among organisms and between organisms and their environments, examining the effects of population size, pollutants, rainfall, temperature, and altitude. Using knowledge of various scientific disciplines, ecologists may collect, study, and report data on the quality of air, food, soil, and water.

Agricultural and food scientists, sometimes referred to as biological scientists, are discussed elsewhere in the site.

Working Conditions

Biological scientists usually work regular hours in offices or laboratories and usually are not exposed to unsafe or unhealthy conditions. Those who work with dangerous organisms or toxic substances in the laboratory must follow strict safety procedures to avoid contamination. Many biological scientists such as botanists, ecologists, and zoologists take field trips that involve strenuous physical activity and primitive living conditions. Biological scientists in the field may work in warm or cold climates, in all kinds of weather. In their research, they may dig, chip with a hammer, scoop with a net, and carry equipment in a backpack. They also may climb, stand, kneel, or dive.

Marine biologists encounter a variety of working conditions. Some marine biologists work in laboratories; others work on research ships. Marine biologists who work underwater must practice safe diving while working around sharp coral reefs and hazardous marine life. Although some marine biologists obtain their specimens from the sea, many still spend a good deal of their time in laboratories and offices, conducting tests, running experiments, recording results, and compiling data.

Some biological scientists depend on grant money to support their research. They may be under pressure to meet deadlines and to conform to rigid grant-writing specifications when preparing proposals to seek new or extended funding.

Employment

Biological scientists held about 77,000 jobs in 2004. Slightly more than half of all biological scientists were employed by Federal, State, and local governments. Federal biological scientists worked mainly for the U.S. Departments of Agriculture, Interior, and Defence and for the National Institutes of Health. Most of the rest worked in scientific research and testing laboratories, the pharmaceutical and medicine manufacturing industry, or hospitals.

In addition, many biological scientists held biology faculty positions in colleges and universities.

Training

A Ph.D. degree usually is necessary for independent research, industrial research, and college teaching, as well as for advancement to administrative positions. A masters degree is sufficient for some jobs in basic research, applied research or product development, management, or inspection; it also may qualify one to work as a research technician or as a teacher in an aquarium. The bachelors degree is adequate for some nonresearch jobs. For example, some graduates with a bachelors degree start as biological scientists in testing and inspection or get jobs related to biological science, such as technical sales or service representatives. In some cases, graduates with a bachelors degree are able to work in a laboratory environment on their own projects, but this is unusual. Some may work as research assistants, whereas others become biological laboratory technicians or, with courses in education, high school biology teachers. Many with a bachelors degree in biology enter medical, dental, veterinary, or other health profession schools.

In addition to required courses in chemistry and biology, undergraduate biological science majors usually study allied disciplines such as mathematics, physics, engineering, and computer science. Computer courses are essential because employers prefer job applicants who are able to apply computer skills to modelling and simulation tasks and to operate computerised laboratory equipment, particularly in emerging fields such as bioinformatics. Those interested in studying the environment also should take courses in environmental studies and become familiar with current legislation and regulations. Prospective biological scientists who hope to work as marine biologists should have at least a bachelors degree in a biological or marine science.

However, students should not overspecialise in undergraduate study, as knowledge of marine biology often is acquired in graduate study. Most colleges and universities offer bachelors degrees in biological science, and many offer advanced degrees. Curriculums for advanced degrees often emphasize a subfield such as microbiology or botany, but not all universities offer all curriculums. Larger universities frequently have separate departments specialising in different areas of biological science. For example, a program in botany might cover agronomy, horticulture, or plant pathology. Advanced degree programs include classroom and fieldwork, laboratory research, and a thesis or dissertation.

Biological scientists with a Ph.D. often take temporary postdoctoral research positions that provide specialised research experience. Postdoctoral positions may offer the opportunity to publish research findings. A solid record of published research is essential in obtaining a permanent position involving basic research, especially for those seeking a permanent college or university faculty position. In private industry, some may become managers or administrators within the field of biology; others leave biology for nontechnical managerial, administrative, or sales jobs.

Biological scientists should be able to work independently or as part of a team and be able to communicate clearly and concisely, both orally and in writing. Those in private industry, especially those who aspire to management or administrative positions, should possess strong business and communication skills and be familiar with regulatory issues and marketing and management techniques. Those doing field research in remote areas must have physical stamina. Biological scientists also must have patience and self-discipline to conduct long and detailed research projects.

Worms can Inherit a 'Memory of Longevity' from Long-lived Parents

Long-lived worms can transmit their extended lifespan to the next generation by passing on changes in the way their genes are used, rather than differences in DNA itself.

A study has shown that nematode worms can inherit a "memory of longevity" from their parents, even though their genome remains unchanged.

It is not clear if the same processes apply to humans, who led the study, noted that some genes that affect the lifespan of nematodes were later found to influence human longevity too. "In several cases, the worm has proved to be a good model for humans, who live 2,000 times longer," she said.

Her team is now looking to see if the results in the worm *Caenorhabditis elegans* translate to species that are evolutionarily closer to humans, such as fish and mice.

The discovery is an example of "epigenetic inheritance", where organisms pass on changes in the way genes are used rather than in the genes themselves. As animals develop, their DNA and proteins become annotated by molecular marks that act like Post-It notes, dictating which genes are read without changing the underlying text. The marks are meant to be stripped away with each new generation, but some stay behind.

Brunet says her study and others show that epigenetic inheritance has broad implications for complex traits like longevity, or possibly even diseases like Alzheimer's, cancer and diabetes. "These have very different causes, including genetic factors and what we do to ourselves. But maybe we're missing a component: what our parents and grandparents have done," she says.

Studies like this could help to improve techniques such as cloning or reprogramming cells to become stem cells, which depend on correctly erasing or maintaining epigenetic marks. Eventually, scientists may even be able to manipulate these marks to boost lifespan or reduce the risk of diseases that pass from one generation to the next.

The latest discovery, published on Wednesday in the journal Nature, began when Brunet's student Eric Greer was working with worms that had mutations in one of three genes – ASH-2, WDR-5 and SET-2. This trio controls how DNA is packaged, converting it from a tightly packed form into a loose, open one where genes can be more easily read.

If any one of the trinity of genes is faulty, various stretches of DNA close up. This restricts access to several genes, including some that are involved in ageing. As a result, the worms live around 20-30% longer.

Greer bred the mutant worms so some of their offspring had working copies of all three genes. But even though their genes were normal, they still lived longer than usual, a trend that continued for three generations. In recent years, scientists have found several examples of epigenetic inheritance, in everything from the symmetry of flowers to the hair colour of mice. But, said Brunet, "We didn't expect that a complex trait like longevity could be inherited in a non-genetic manner."

In Greer's study, the fact that the long life disappears after three generations makes it very unlikely that genetic changes are involved. "This will stimulate researchers to take the notion more seriously," said Whitelaw.

Epigenetic inheritance is often portrayed as a revolution in evolution. It seems to hark back to the discredited ideas of Jean Baptiste Lamarck, who suggested that traits acquired by parents would be transmitted across the generations.

"The discovery that epigenetic changes can carry across generations has made people realise that not everything is written in the genes," said Brunet. "That's not to say that Lamarck was right, but we now understand that there's epigenetic memory for some traits."

Epigenetic inheritance could give organisms a short-term advantage, by giving the next generation advance notice of the challenges that the current one is facing. "If there's a famine, it might be advantageous to have this information passed on," said Brunet.

The long-term consequences of this are not clear. The long life of Greer's worms, like most other examples of epigenetic inheritance, disappears after a few generations. They are unlikely to have a lasting impact on the fate of the species. "It's not a long-term evolutionary thing," said Brunet.

Cell Biology

Cell biologists study the physiological properties, behaviour, interactions, and environment of cells, at the microscopic and molecular levels. That broad job description covers a wide variety of individual activities. "For basic researchers, the key areas are the very fundamental processes that cells undergo. They study adhesion, differentiation, apoptosis, migration, and many specialised functions of specific cell types," says Laurel Donahue, manager of technical development at Sigma Aldrich. "Industrial cell biologists have narrower focus. They look at the underlying mechanisms that improve the longevity of cells in culture, as well as other processes that will improve protein production. They look at processes that will produce consistent cultures of cells at large scale, and they focus a lot on cell based assays for identifying targets and screening compounds for pharmacological activity."

Whether they work in research or development, most modern cell biologists have one key goal in mind. "The main theme is gaining a better understanding of diseases, through good statistical correlation of molecular interactions with phenotypes," says Ulrich Simon, head of the microscopy business group at Carl Zeiss. "We try to shine more light into understanding the root causes of diseases."

That effort divides naturally into specific project areas. For example, the diversity of signal transduction pathways has recently emerged as a critical focus for researchers. "When cells interact with the extracellular milieu, they have to be able to activate the pathways," says Erik Schaefer, vice president for R&D in the signal transduction area at Biosource International, a company recently acquired by Invitrogen. "Understanding that is the area of functional genomics and proteomics."

Stem Cells and Beyond

J±rg (george) Pochert, director of the pharma and biotech group at Hamilton Life Science Robotics, highlights another area of frantic activity: stem cells. "Scientists face issues of how to do stem cell culturing right," he says. "Also related to stem cells are the questions of what kinds of differentiation protocols are available and how good and stable they are, to ensure that scientists can start a stem cell line and differentiate it into the kind of tissue they need."

To achieve their goals, cell biologists use a mixture of traditional and new tools and technologies, from cell cultures to microscopes to protein microarrays. "Recent techniques include higher resolution mass spectrometry and developments involving very good antibodies," says Mike Schutkowski, vice president of R&D and operations for JPT Peptide Technologies. Donahue points out other approaches that have emerged in cell biology labs. "Short interfering RNA continues to be a very important tool," she says. "In addition, information from genomic and proteomic profiling is leading to more hypotheses that can be tested. Industrially, in silico modelling from such companies as Genomatica allows for predictive biology. And on the horizon is nanotechnology."

Laboratory automation has also found a critical role in the cell biology lab. "Every major pharma and all the major biotechnology companies have a great deal of automation in both their discovery and manufacturing sites," explains Doug Gurevitch, a senior development engineer at the University of California, San Diego who is program chair for the Association for Laboratory Automation.

Whatever tools they use, cell biologists face a critical problem common to many other life scientists. "The key factor we see over and over again is trying to bring together the data from all the tools out there," explains Brendan Yee, strategic product manager at Beckman Coulter. "Cell biologists, especially in drug development, are running into a roadblock created by the amount of data they collect."

The effort to overcome that and other problems in pursuing cell biology involves R&D scientists in academe and industry. "you'll involved in the Cell Migration Consortium," Schaefer says. "This has multimillion dollar projects aimed at bringing together large groups of scientists and making discoveries commercially available. we're working with the consortium to commercialise tools."

Medieval Methodologies

The study of any living cell starts with the use of cell culture media and reagents. "Cell culture media, sera, and related reagents are critically important - more than even investigators who have done cell culture for a long time think," Donahue says. "Culture systems are downright medieval at times." To bring cell cultures into the 21st century and keep cells alive and well during in vitro experiments, researchers and manufacturers have developed a range of growth media, some of which contain only well-defined or synthetic components. Scientists can supplement these defined or 'serum-free media with growth factors to study the living cells' responses to changing environments.

Companies such as ATCC, Cambrex, Invitrogen, and Sigma-Aldrich offer cell culture media, sera, and reagents to grow cells ranging from HeLa to embryonic stem cells. "We look at systems that are being utilised by researchers and try to create products and culture systems that are, for example, serum-free or better defined," Donahue says.

"For our industrial customers, who have regulatory constraints, we focus on the development of animal derived component free media." Recent products from Sigma-Aldrich include Excel Vero, a medium for cells used in research on or production of viral vaccines and two kits that permit industrial researchers to optimise the media for their protein expressing CHO cell lines.

Automation can play a particularly effective role in cell cultures. "Bulk cell culture has been automated for years," Gurevitch explains. "We've been pushing newer work with microcultures for quick, tailored cell development for research, development, and eventually tissue engineering."

Hamilton recently created what it calls its MICROLAB STAR based Cell-Host System. Developed in conjunction with Oliver Bruestle of Germany's University of Bonn, this automates several aspects of stem cell culture, including media exchange, cell harvesting after trypsinisation, plating cells, and the addition of growth factor or other substances to the cell cultures. "The system offers full weekend walk-away times for lab personnel," Pochert says. "It ensures that all the cells are treated homogeneously and that the biological bandwidth of cells you are culturing is narrower than if you use manual methods. We have used it for mouse stem cells, on which we have published two papers."

The Importance of Imaging

Whenever possible, cell biologists want their experiments to show rather than tell. Tools developed in recent years have facilitated that. One area of novel technology overcomes the inherent limitation that cells are colourless and translucent, and hence almost invisible under the lens of a standard light microscope. Biological stains and dyes enable microscopists to look at different cell types and the structures within these cells. "We want to use dyes to indicate how cells interact," Carl Zeiss's Simon explains. Companies such as Fisher Scientific, USB Corporation, and Wako Chemicals offer several stains and dyes, many of them highly purified and use-tested to ensure consistent performance in the laboratory.

Other methods for improving the visibility of cells complement microscopy methods. "Spectral deconvolution techniques introduced in confocal and microscopy techniques put the colour issue under control," Simon says. "It has added value to live imaging." Zeiss has recently introduced LSM5 Live, a laser scanning microscope developed in collaboration with Scott Fraser of the California Institute of Technology. "This speeds up visualisation by a factor of 20 to 30," Simon says.

"That opens up new dimensions on research; you can understand colour and you're capable of really understanding how proteins interact." In addition to Zeiss, major producers of microscopes include Leica, Nikon Instruments, and Olympus. Beyond refining microscopes, those vendors have developed digital camera systems and software for data analysis.

Apogee/Biodimensions, meanwhile, has developed two noninvasive technologies for in vivo functional imaging at the molecular level. One monitors epithelial inflammation and the other recognises concentrations of analysts that indicate disease or treatment status. The technology relies on the company's Single System Image (SSI) high performance computing cluster. It permits scientists to compare results from image processing algorithms and photon simulation codes, and thereby evaluate the performances of the approaches during development and testing.

Assays and Antibodies

Imaging doesn't work in isolation. Cell scientists also rely on biochemical assays to determine cellular function. Well-studied subjects of assaying include apoptosis, the cell's cytoskeleton and extracellular matrix, protein phosphorylation, signal transduction pathways, ion channels, nitric oxide, and cell stress. Early entrants into assaying for cell biology include Alexis Biochemicals, Biosource, and Tocris Cookson. Apoptosis, or programmed cell death, has particular traction. Those companies and others such as EMD Biosciences, Roche Applied Science, and Trevigen offer assays for studying it.

Antibodies represent major research tools for cell biology and signal transduction. For example, a key development in the study of kinases - molecules that control many cellular functions - was the production of specific targeted antibodies that identify the active form of protein kinases. Those tools improved researchers' understanding of the role of kinases in various signal transduction pathways. Companies such as Biosource, BD Biosciences Pharmingen, Chemicon, and Upstate offer antibodies to study the role of kinases in signal transduction.

BioSource has another focus. "One of the things we've done with our antibodies is put them into platforms to understand the kinetics and overall stoichiometry of phosphorylation events," Schaefer explains. "We have patent-pending technology to detect phosphorylation for certain protein events and to work out how much is phosphorylated. We also have a phosphoarray chip platform. And we offer beads that allow you to get up to 100 combinations and dye signatures to capture proteins of interest, and identify them with a detection antibody."

In September, Beckman Coulter introduced a custom kit to measure ZAP-70 protein expression in whole blood specimens using flow cytometry. ZAP-70 (Zeta-associated protein, 70 kilodaltons) is an intracellular protein associated with signal transduction networks in lymphocyte populations whose expression has been associated with disease progression in a group of patients with chronic lymphocytic leukemia, the most common leukemia in the United States and Europe.

"There's interest in whether ZAP-70 protein expression marks a group of patients in whom the disease will progress rapidly," explains Vincent Shankey, a senior staff development scientist at Beckman Coulter's Advanced Technology Centre. "Our kit contains an optimised antibody to ZAP-70, a unique fixation and permeabilisation technique to monitor intracellular signal transduction protein epitopes, a gating and analysis technique to measure ZAP-70 protein expression in key cell populations within the sample, and a rigorous protocol for researchers to follow to set up a flow cytometre to measure the protein."

Microarraying Methods

The introduction of DNA microarrays quickly sparked the idea of using this miniaturised platform with other biomolecules, such as proteins and peptides. Biacore and Ciphergen have developed proprietary array systems for analysing proteins. And JPT Peptide Technologies has developed its own approach. We were first on the market with peptide arrays in 2000," Schutkowski says. "We have technology that synthesizes thousands of peptides per week."

The firm's unique peptide microarrays accelerate crucial steps in drug discovery programs by focusing on enzyme families such as kinases and proteases. The company's pepSTAR platform identifies substrates in less than one week and profiles kinases and proteases.

JPT also customises peptide chips to target specific classes of enzymes. And it provides an all-in-one service that includes screening microarrays, evaluating data, and synthesizing, characterising, and optimising selected substrates. ±The logical extension of these products," Schutkowski explains, "is to use the microarrays with samples from human patients - looking for antibodies against HIV, for example."

Advances in cell biology hold great promise for basic scientific understanding and the treatment of disease. The tools and techniques in application and under development offer the promise of discoveries that will add to scientists' understanding of how cells function and the effective application of that knowledge.

Advances in Biology and Therapy of Multiple Myeloma

Even during this past year, further advances have been made in understanding the molecular genetics of the disease, the mechanisms involved in the generation of myeloma-associated bone disease and elucidation of critical signalling pathways as therapeutic targets. New agents (thalidomide, Revimid, Velcade) providing effective salvage therapy for end-stage myeloma, have broadened the therapeutic armamentarium markedly.

As evidenced in Section I by Drs. Kuehl and Bergsagel, five recurrent primary translocations resulting from errors in IgH switch recombination during B-cell development in germinal centres involve 11q13 (cyclin D1), 4p16.3 (FGFR3 and MMSET), 6p21 (cyclin D3), 16q23 (c-maf), and 20q11 (mafB), which account for about 40% of all myeloma tumours.

Based on gene expression profiling data from two laboratories, the authors propose 5 multiple myeloma (MM) subtypes defined by the expression of translocation oncogenes and cyclins (TC molecular classification of MM) with different prognostic implications. In Section II, Drs. Barillé-Nion and Bataille review new insights into osteoclast activation through the RANK Ligand/OPG and MIP-1 chemokine axes and osteoblast inactivation in the context of recent data on DKK1.

The observation that myeloma cells enhance the formation of osteoclasts whose activity or products, in turn, are essential for the survival and growth of myeloma cells forms the basis for a new treatment paradigm aimed at reducing the RANKL/OPG ratio by treatment with RANKL inhibitors and/or MIP inhibitors.

In Section III, Dr. Fenton reviews apoptotic pathways as they relate to MM therapy. Defects in the mitochrondrial intrinsic pathway result from imbalances in expression levels of Bcl-2, Bcl-XL and Mcl-1. Mcl-1 is a candidate target gene for rapid induction of apoptosis by flavouperidol. Antisense oglionucleotides (ASO) lead to the rapid induction of caspace activity and apoptosis, which was potentiated by dexamethasone. Similar clinical trials with Bcl-2 ASO molecules alone and in combination with doxorubicin and dexamethasone or thalidomide showed promising results.

The extrinsic pathway can be activated upon binding of the ligand TRAIL. OPG, released by osteoblasts and other stromal cells, can act as a decoy receptor for TRAIL, thereby blocking its apoptosis-inducing activity. MM cells inhibit OPG release by stromal cells, thereby promoting osteoclast activation and lytic bone disease (by enhancing RANKL availability) while at the same time exposing themselves to higher levels of ambient TRAIL. Thus, as a recurring theme, the relative levels of pro- versus anti-apoptotic molecules that act in a cell autonomous manner or in the milieu of the bone marrow microenvironment determine the outcome of potentially lethal signals.

In Section IV, Dr. Barlogie and colleagues review data on single and tandem autotransplants for newly diagnosed myeloma. CR rates of 60%–70% can be reached with tandem transplants extending median survival to ~7 years. Dose adjustments of melphalan in the setting of renal failure and age > 70 may be required to reduce mucositis and other toxicities in such patients, especially in the context of amyloidosis with cardiac involvement.

In Total Therapy II the Arkansas group is evaluating the role of added thalidomide in a randomised trial design. While data are still blinded as to the contribution of thalidomide, the overriding adverse importance of cytogenetic abnormalities, previously reported for Total Therapy I, also pertain to this successor trial. In these two-thirds of patients without cytogenetic abnormalities, Total Therapy II effected a doubling of the 4-year EFS estimate from 37% to 75% ($P < .0001$) and increased the 4 year OS estimate from 63% to 84% ($P = .0009$).

The well-documented graft-vs-MM effect of allotransplants can be more safely examined in the context of non-myeloablative regimens, applied as consolidation after a single autologous transplant with melphalan 200 mg/m^2, have been found to be much better tolerated than standard myeloablative conditioning regimens and yielding promising results even in the high-risk entity of MM with cytogenetic abnormalities.

For previously treated patients, the thalidomide congener Revimid and the proteasome inhibitor Velcade both are active in advanced and refractory MM (~30% PR).

Gene expression profiling (GEP) has unraveled distinct MM subtypes with different response and survival expectations, can distinguish the presence of or future development of bone disease, and, through serial investigations, can elucidate mechanisms of actions of new agents also in the context of the bone marrow microenvironment. By providing prognostically relevant distinction of MM subgroups, GEP should aid in the development of individualised treatment for MM.

Chapter 4

Molecular Genetics for Disease Classification and Identification

Multiple myeloma (MM), currently an incurable malignancy that often is preceded by premalignant monoclonal gammopathy of undetermined significance (MGUS), has a yearly incidence of nearly 14,000 in the US. For both MGUS and MM, the incidence is markedly age dependent, about 2-fold higher in American blacks than Caucasians, and significantly higher in males. The roles of genetic background and environment are poorly defined, although there may be clustering within families.

MM Is a Plasmablast/Plasma Cell Tumour of Postgerminal Centre B Cells

Most B cell tumours, including MM, involve germinal centre (GC) or post-GC B cells. Germinal centre B cells uniquely modify their DNA through sequential rounds of somatic hypermutation and antigen selection, and also by immunoglobulin heavy chain (IgH) switch recombination. Post-GC B cells can generate plasmablasts (PB) that have successfully completed somatic hypermutation and IgH switching before migrating to the bone marrow (BM), where stromal cells enable terminal differentiation into long-lived plasma cells (PC). Although PC can be generated from either pre-GC or post-GC B cells, premalignant nonIgM MGUS and malignant MM are post-GC clonal tumours with phenotypic features of PB/PC, and are distributed at multiple sites in the bone marrow. A critical feature shared by MGUS and MM is the presence of a substantial tumour mass despite an extremely low rate of proliferation, usually with less than 1% of tumour cells synthesizing DNA until late stages of MM.

Stages of Multiple Myeloma

There are a number of clinically defined stages for MM tumours. A clonal PC neoplasm must expand to about 10 cells before it produces enough immunoglobulin to be recognised as a monoclonal Ig "spike" (M-Ig) by serum electrophoresis. For premalignant MGUS, which typically is asymptomatic and stable, the M-Ig is d" 3 g/dl and the tumour cells comprise no more than 10% of the mononuclear cells in the BM. However, depending on the level of M-Ig (a surrogate measure of the number of tumour cells), 0.6%–3% per year of patients with non-IgM MGUS progress to MM expressing the same M-Ig. At present, there are no unequivocal genetic or phenotypic markers that distinguish MGUS from MM tumour cells, so it is not possible to predict if and when a particular MGUS tumour will progress to MM.

Also, it remains unclear to what extent intrinsic genetic or epigenetic changes in the MGUS tumour cell versus extrinsic changes in nontumour cells (e.g., immune cells) are responsible for this progression. *Primary amyloidosis*, which accounts for about 4000 deaths per year in the United States, appears to represent premalignant MGUS that is symptomatic because of pathological deposits of the M protein (generally the intact or fragmented Ig light chain) in various tissues. MM is distinguished from MGUS by having a BM tumour cell content of > 10%. *Smoldering* MM (SMM), which has a stable intramedullary tumour cell content of > 10% but no osteolytic lesions or other complications of malignant MM, has a high probability of progressing to frankly malignant MM, which is distinguished by having osteolytic bone lesions and/or an increasing tumour mass.

Further progression of MM is associated with increasingly severe secondary features (lytic bone lesions, anemia, immunodeficiency, renal impairment), and in a fraction of patients, the occurrence of tumour in extramedullary locations. Extramedullary MM is a more aggressive tumour that often is called secondary or primary plasma cell leukemia, depending on whether preceding intramedullary myeloma has been recognised. Human MM cell lines (HMCL) can sometimes be generated, but usually only from extramedullary tumours.

Oncogene Dysregulation by Ig Translocation: A Hallmark of B Cell Tumours

A seminal event in most kinds of B cell tumours (chronic lymphocytic leukemia being a major exception) is dysregulation of an oncogene that is juxtaposed near a strong Ig enhancer as a result of translocation to the IgH locus (14q32), or (less often) to an IgL locus. These translocations appear to be mediated mainly by errors in 1 of 3 B cell specific DNA modification mechanisms: V(D)J recombination that occurs during early B cell development, IgH switch recombination, and somatic hypermutation, the latter 2 processes occurring mainly in germinal centre B cells. The mechanism(s) by which double-stranded DNA breaks are generated in the partner chromosome are poorly understood. However, for many kinds of B cell tumours, there is a consistent Ig translocation that involves only 1 major partner.

Translocations Involving Either an IgH or IgL Locus are Frequent in Myeloma

The combination of karyotypic complexity, an inability to efficiently perform conventional cytogenetics on low proliferative tumours, and the telomeric location of some translocation partners delayed the identification of Ig translocations in MGUS and MM. An important initial step in solving this problem was the identification and cloning of IgH translocation breakpoints in HMCL. The subsequent application of interphase fluorescence in situ hybridisation (FISH) using specific genetic probes to identify karyotypic abnormalities even in nondividing cells that have a PC phenotype enabled analysis of primary MGUS and MM tumours.11,12 Many recent studies have shown that many MM tumours have an IgH translocation that nonrandomly involves one of many potential chromosomal partners (for example, Bergsagel and Kuehl, Avet-Loiseau et al, and Fonseca et al).

The prevalence of IgH translocations varies with the stage of disease: 46%–48% in MGUS or SMM, 55%–73% in intramedullary MM, 85% in primary plasma cell leukemia, and > 90% in HMCL. Limited studies indicate a much lower prevalence of Igë translocations: 11% in MGUS, 17% in advanced MM tumours, and 23% in HMCL. The prevalence of Igê translocations is even lower, and appears to be no more than 2%–3% based on studies of advanced MM tumours and HMCL. Importantly, although all 34 HMCL fully analysed have either an IgH or Igë translocation, nearly 50% of MGUS tumours and at least 26% of advanced MM tumours have neither an IgH nor an IgL translocation.

Recurrent Chromosomal Partners for Ig Translocations

Apart from c-myc at 8q24, there are 5 well-defined recurrent chromosomal partners (oncogenes) that are involved in IgH translocations in MGUS and MM: 11q13 (cyclin D1), 4p16.3 (FGFR3 & MMSET), 6p21 (cyclin D3), 16q23 (c-maf), and 20q11 (mafB), with at least the latter 3 also involved in Igë translocations. Together the combined prevalence of these 5 IgH translocation partners is about

40%, with approximately 15% 11q13, 15% 4p16, 3% 6p21, 5% 16q23, and 2% 20q11., The t(4;14) translocation is unusual in that it appears to dysregulate 2 potential oncogenes, FGFR3 on der (14) and MMSET on der (4), but FGFR3 on der (14) is lost or not expressed in about 20% of MM tumours that have a t(4;14) translocation., There is conflicting data regarding the issue of whether the prevalence of t(4;14) and t(14;16) translocations is the same or much lower in MGUS/SMM compared to MM. However, the apparently lower incidence of 4p16 and/or 16q23 in MGUS/SMM compared to MM may be due to these translocations resulting in de novo MM without preceding MGUS, or a more rapid progression of MGUS to MM, an hypothesis supported by the fact that patients with translocations involving 4p16 or 16q23 have an extremely poor prognosis.

Primary Versus Secondary Translocations in MM

Primary translocations occur as early and perhaps initiating events during tumour pathogenesis, whereas secondary translocations occur as progression events. For B cell tumours, most primary translocations appear to be simple reciprocal translocations that juxtapose a partner chromosomal locus (and oncogene) and one of the Ig enhancers, and usually are mediated by 1 of the 3 B-cell–specific DNA modification mechanisms described above. For MM, most translocations involving the 5 recurrent translocation partners described above appear to be primary translocations that occurred from errors in IgH switch recombination (possibly less often errors in somatic hypermutation) during B cell development in germinal centres. Occasionally, however, it appears that secondary translocations might involve 1 or more of these 5 recurrent partners, (C Cultraro and A Gabrea, unpublished).

In addition, there are rare tumours that have independent translocations involving 2 of these 5 recurrent partners; all combinations have been documented except translocations that involve both 6p21 and 11q13 (C Cultraro and A Gabrea, unpublished). For normal PC and PC tumours, it appears that B-cell–specific DNA modification mechanisms are inactive. Therefore, unless one of these mechanisms could be reactivated, secondary translocations in PC tumours would be mediated by other kinds of recombination mechanisms that do not specifically target the Ig loci but could involve an Ig locus. In contrast to primary translocations, secondary translocations usually are complex, unbalanced translocations or insertions, often involving 3 different chromosomes and sometimes with associated inversion, deletion, duplication, or amplification.

Primary translocations should be present in all tumour cells in both MGUS and MM, whereas secondary translocations are expected to be less frequent in MGUS than in MM, and might be present in only a subset of MGUS or MM tumour cells. Obviously, however, the only definitive way to distinguish primary from secondary translocations would be to document the time(s) at which translocations occur during the progression of individual tumours. In the absence of this definitive test, the criteria described above provide some help in distinguishing primary from secondary translocations.

Dysregulation of Myc: A Paradigm for Late Secondary Translocations in MM

Similar to other kinds of B cell tumours, translocations that dysregulate c-myc represent an important pathogenic event in MM. Chromosomal translocations that dysregulate c-myc by juxtaposing it with one of the three Ig loci represent an essentially invariant and apparently primary event in human Burkitt's lymphoma and murine plasmacytoma tumours. The nontranslocated c-myc allele is not expressed, corresponding to the absence of c-myc expression, in resting germinal centre B cells and terminally differentiated plasma cells. Strikingly, L-myc (1 HMCL) or 1 c-myc allele is expressed selectively in all informative HMCL, consistent with *cis*-dysregulation of L-myc or 1 c-myc allele in all HMCL. In addition, by our analysis of published gene expression profiling, N-myc (which is not expressed in normal PB or PC) is expressed in 2 of 82 primary MM tumours.

Three colour FISH analyses of metaphase chromosomes show that nearly 90% of HMCL and 50% of advanced MM tumours have similar karyotypic abnormalities involving c-myc, L-myc (1 HMCL), or N-myc (1 tumour). Simple, reciprocal t(8;14) and t(8;22) translocations are infrequent. Most karyotypic abnormalities are complex translocations and insertions that often are nonreciprocal, and frequently involve 3 different chromosomes. Karyotypic abnormalities involving c-, L-, or N-myc often do not include association with an Ig enhancer, which suggests that secondary translocations can dysregulate c-myc by juxtaposition to non-Ig enhancers.

By interphase FISH analyses, it is reported that the c-myc locus is rearranged in 3% of MGUS/SMM tumours, 10% of MM tumours with a low tumour mass, and 19% of MM tumours with a high tumour mass (â2 microglobulin > 3), and frequently is heterogeneous within a tumour. Cloned t(8;14) translocation/insertion breakpoints usually do not occur at the IgH sites targeted by the 3 B-cell–specific DNA modifications. All of these results support a model for MM in which dysregulation of c-, L-, or N-myc occurs as a late progression event that is mediated by secondary translocations not involving the 3 B-cell–specific DNA modification mechanisms.

Promiscuous Partners for Ig Translocations

Interphase FISH studies suggest that approximately 20% of MM tumours have IgH translocations not involving 1 of the recurrent loci cited above. Similarly, spectral karyotypic (SKY) analyses of metaphase chromosomes from 150 advanced MM tumours show that 15% of tumours have IgH translocations that do not involve a myc gene or one of the five recurrent partners described above. Most of these novel chromosomal loci are involved in only one MM tumour, and none are involved in more than three of the 150 MM tumours analysed. It is unclear to what extent these promiscuous partners are generated by primary translocations versus secondary translocations.

However, it is notable that IgH translocations involving the recurrent 5 chromosomal loci are found mainly in nonhyperdiploid tumours, whereas translocations involving other partners are represented to a similar extent in hyperdiploid and nonhyperdiploid tumours. More importantly, the identity and significance of the promiscuous partners (with the exception of IRF-4 at 6p25) remain enigmatic.

Dysregulation of Cyclin D1, 2, or 3: A Possible Unifying Oncogenic Event in MM

In terms of proliferation, MGUS and MM seem closer to normal, nonproliferating PC than to normal, but highly proliferating PB, for which 30% or more of the cells can be in S phase. Surprisingly, however, our analysis of combined gene expression profiling data from two laboratories shows that the expression level of cyclin D1, cyclin D2, or cyclin D3 mRNA in MM (and MGUS based on analysis of a limited number of samples) is relatively high, comparable to the levels of cyclin D2 mRNA expressed in normal proliferating PB, and distinctly different from normal PC. Normal hematopoietic cells, including normal B lymphocytes and PB, predominantly express cyclin D2, usually together with lower levels of cyclin D3, but do not express cyclin D1 (reviewed in Shaughnessy et al).

Given the lack of cyclin D1 expression in normal lymphocytes and the occurrence of Ig translocations that dysregulate cyclin D1 or cyclin D3 in a subset of MM tumours it seems apparent that virtually all

MM tumours dysregulate at least 1 of the cyclin D genes. From cyclin D1 transgenic mice and cyclin D transfection experiments, it is known that overexpression of cyclin D is insufficient by itself to cause cell cycle progression. Instead, it has been suggested that the overexpression of cyclin D renders a cell more sensitive to growth activating signals and/or less sensitive to growth inhibitory signals. The apparent universal enhanced expression/dysregulation of one of the cyclin D genes in low proliferative MM tumours seems consistent with what is known about the effect of dysregulated cyclin D expression in these model systems.

Five MM Subtypes Defined by Expression of Translocation Oncogenes and Cyclin D

In addition to determining the expression level of cyclin D1, 2, and 3, gene expression profiling can effectively identify MM tumours that overexpress the oncogenes dysregulated by the 5 recurrent IgH translocations: 11q13 (cyclin D1); 6p21 (cyclin D3); 4p16 (MMSET and usually FGFR3); 16q23 (c-maf); and 20q11 (mafB) (and unpublished Affymetrix Hu95Av2 and Hu133A+B data from J. Shaughnessy lab). Unsupervised hierarchical cluster analysis of microarray gene expression profiles (Affymetrix Hu95Av2) confirms that there are a minimum of 5 distinct groups, with only limited overlap of the different groups.

These groups can be distinguished based on the Ig translocation present, and cyclin D expression (TC classification): Group TC1 tumours (18%) express high levels of either cyclin D1 or cyclin D3 as a result of an Ig translocation; Group TC2 tumours (43%) ectopically express low to moderate levels of cyclin D1 despite the absence of a t(11;14) translocation; Group TC3 tumours (17%) are a mixture, with most expressing cyclin D2, but a few expressing only very low levels of cyclin D2 and/or cyclin D3; Group TC4 tumours (15%) express high levels of cyclin D2, and also MMSET (and in most cases FGFR3) as a result of a t(4;14) translocation; Group TC5 tumours (7%) express the highest levels of cyclin D2, and also high levels of either c-maf or mafB, consistent with the possibility that both maf transcription factors up-regulate the expression of cyclin D2.

Other Thoughts about the Roles of Ig Translocations and Cyclin D Dysregulation in MM

As indicated above, the incidence of IgH translocations is correlated with the stage of the disease. The occurrence of 2 (or sometimes 3) independent IgH translocations is even more prominently correlated with the stage of disease. In addition, the prevalence of the 5 recurrent IgH translocations (particularly 16q23, and 20q11, but also 4p16 and 11q13) and c-myc translocations is lower in intramedullary tumours than in HMCL that are derived from extramedullary tumours representing both primary PC leukemia and terminal progression of intramedullary MM. The increased incidence of secondary c-myc translocations in HMCL is consistent with accumulation of these translocations during disease progression.

However, the increased incidence of primary translocations in HMCL compared to intramedullary tumours most likely occurs as a result of selective progression of intramedullary tumours with Ig translocations to an extramedullary phase from which virtually all HMCL are generated. Consistent with this latter hypothesis, Group TC2 tumours that ectopically express cyclin D1 without a t(11;14) translocation are not represented among the 34 HMCL that we have analysed. On the basis of these results, we suggest the hypothesis that progression from stromal dependent, intramedullary MM to stromal independent, extramedullary MM requires a minimum of 2 genetic events, 1 of which can be an Ig translocation that dysregulates expression of one of the cyclin D genes As a corollary to this hypothesis, we suggest that the ectopic expression of cyclin D1 in the group TC2 MM tumours requires the interaction of the MM tumour cells with bone marrow stromal cells.

Two pathways of generating monoclonal gammopathy of undetermined significance (MGUS)/multiple myeloma (MM) tumours.

Tumours can be generated either with (+) or without (") the occurrence of a primary translocation, the former being mainly nonhyperdiploid and the latter mainly hyperdiploid. Dysregulation of a cyclin D gene appears to be an early event in each case. Secondary translocations can occur in both kinds of tumours. As indicated, tumours in the former pathway selectively generate stromal independent MM cell lines, but the dashed arrows suggest the possibility that the latter pathway might generate cell lines.

A Current Model for the Molecular Pathogenesis of Multiple Myeloma

Considering the results summarised above, we propose the following model for the molecular pathogenesis of MM. In at least 40%—but possibly up to 60%—of tumours, a primary chromosome translocation (mediated mostly by aberrant switch recombination, and less frequently by aberrant somatic mutation), or possibly an early secondary translocation, results in the ectopic expression of an oncogene. This may lead directly (11q13–cyclin D1 and 6p21–cyclin D3) or indirectly (4p16, 16q23, other–cyclin D2) to cyclin D dysregulation. Alternatively, in at least 40%—but possibly up to 60%—of tumours there is no primary translocation, and cyclin D1 (possibly sometimes cyclin D2) is dysregulated by an as yet undefined mechanism that may involve aberrant interaction with bone marrow stromal cells.

The dysregulation of 1 of 3 cyclin D genes, which provides a unifying model for the pathogenesis of MM, may render these clonal cells more susceptible to proliferative stimuli, resulting in selective expansion of this clone as a result of interaction with bone marrow stromal cells that produce interleukin (IL)-6 and other cytokines.

Numeric (and possibly structural) karyotypic abnormalities, most notably including trisomies of chromosomes 3, 5, 7, 9, 11, 15, 19, and 21, and monosomy of chromosome 13 or deletion of 13q14, often are present in premalignant MGUS, the earliest identified stage of tumourigenesis. It remains to be determined if karyotypic abnormalities occur before or after primary IgH translocations. However, it is noteworthy that monosomy of chromosome 13 (or 13q14 deletion), which is present in approximately 50% of MM tumours.11,12 is present in most MM tumours that have a t(4;14) or t(14;16) translocation, perhaps consistent with the occurrence of this karyotypic abnormality preceding these translocations.

It also is of interest that hyperdiploid tumours, which have heterogeneous combinations of trisomies involving the odd number chromosomes cited above, have a very low incidence of the 5 recurrent translocations compared to nonhyperdiploid tumours.

It would appear that the hyperdiploid tumours are enriched for the group TC2 tumours that ectopically express cyclin D1 but probably have only infrequent IgH translocations. It is not clear if there is a high rate of ongoing karyotypic instability in MM, but tumour progression is associated with secondary chromosome translocations, of which c-myc provides a paradigm. The secondary translocations of c-, N-, or L-myc, which are associated with mono-allelic expression of the corresponding myc gene, are present at a low frequency in intramedullary MM tumours, but occur in nearly 50% of advanced tumours that generate metaphase chromosomes, and are almost universally present in HMCL. By timing, this dysregulation of a *myc* gene would appear to be associated with progression to a more aggressive, proliferative phenotype, perhaps decreasing the requirement for stromal cell cytokines that stimulate c-myc expression and proliferation at earlier disease stages. Mutually exclusive activating mutations of K- or N-Ras (or FGFR3 when there is a t(4;14) translocation) are rare or absent in MGUS, whereas RAS mutations are present in 30%–40% of early MM but perhaps a somewhat higher fraction of advanced MM, and the FGFR3 mutations are more frequent in the advanced stages of MM.

Mutations and/or mono-allelic deletion of p53 are seen late in the course of the disease. Interestingly, although dysregulation of a cyclin D gene appears to be a nearly universal event in the pathogenesis of MM, inactivation of Rb or the INK4 cyclin dependent kinase inhibitors still occurs, but perhaps late in the more aggressive phase of the disease: p16INK4a by methylation, and Rb or p18INK4a by bi-allelic deletions.

Molecular Phenotypes Predict Prognosis and Response to Existing Therapies

In addition to tumour mass and secondary features that represent a host response to malignant MM (anemia, bone disease, immunodeficiency, etc.), intrinsic properties of the tumour cell are also informative in predicting prognosis and response to existing therapies. For example, it has been well documented that an unfavourable outcome is associated with each of the following: increased plasma cell labelling index, the generation of tumour cells with an abnormal karyotype (perhaps a surrogate for increased proliferation), hypodiploidy compared to hyperdiploidy, monosomy of chromosome 13/13q, and monosomy of chromosome 17/deletion of p53. It also has been reported but not confirmed that activating mutations of K-Ras (but not N-Ras) represent an adverse prognostic factor.

More recently, it has become clear that specific IgH translocations, which represent early if not initiating events in tumourigenesis, also have a profound prognostic significance. In particular, patients with tumours that have a t(4;14) translocation (TC4) have a substantially shortened survival with either standard or intense therapy, and patients with a t(14;16) (TC5) have a similarly poor if not worse prognosis. By contrast, patients with tumours that have a t(11;14) translocation (TC1) appear to have a marginally better survival following conventional chemotherapy but apparently a remarkably better response to intense therapy. These results suggest that the TC classification, based on translocation and cyclin D expression, and supported by hierarchical cluster analysis, may be a clinically useful way to classify patients into groups that have distinct subtypes of MM (and MGUS) tumours. In practice it would be very hard to get sufficient patients with either 6p21 or 20q11 translocations for a meaningful analysis.

Therefore, given the similar pathogenesis and shared pattern of gene expression for tumours with 11q13 and 6p21 (TC1) and tumours with 16q23 and 20q11 (TC5) we suggest that in each case the patients be considered as part of the same group.

Identification of Novel Therapeutic Strategies

The critical role of cyclin D dysregulation in the pathogenesis of MM highlights the importance of the cyclin D/RB pathway, and suggests that there may be a therapeutic window in targeting this pathway for all molecular subtypes of MM. For example, epigenetic silencing of CDK inhibitor (INK4A) mRNA expression might be reversed by histone deacetylase inhibitors (SAHA, depsipeptide), or inhibitors of DNA methyl transferase (5 aza-22 deoxy-cytidine). To target cyclin D per se, there are a number of possible strategies including modulation of mRNA translation, post-translational modifications, enzyme function, and perhaps even expression of cyclin D mRNA.

First, the cyclin D mRNA is under strict translational control, and agents that inhibit the translation have been identified (e.g., desferroxamine, eicosapentaenoic acid). Second, cyclin D and many other cell cycle regulated proteins are post-translationally regulated by ubiquitination and proteasomal degradation, which might provide another therapeutic target. Third, important steps downstream of cyclin D are dependent on kinases (CDKs) that may be targeted by selective kinase inhibitors.

Finally, as suggested above, it is particularly intriguing that the TC2 group, which includes nearly half of MM tumours, may be absolutely dependent on an interaction with BM stromal cells for the ectopic expression of cyclin D1 that appears to be a critical property for these tumour cells. Clearly we need to elucidate the mechanism that is responsible for the ectopic expression of cyclin D1 in the TC2 group, but there may already be data with new emerging therapies that will help to address this issue. Recently, there have been promising therapeutic results with both thalidomide and its derivatives, and also bortezomib (PS-341), a proteasome inhibitor. Both kinds of treatment appear to target the MM tumour cells but also the interaction of tumour cells with the bone marrow microenvironment, and in each case an as yet undefined subset of patients may be more responsive. It is possible, for example, that the TC2 subgroup is selectively targeted. In any case, for this and for other treatments, it obviously is critically important to determine the response among the different TC groups, and to study the changes in protein levels of cell cycle proteins induced by treatment.

Additional specificity may be achieved by targeting the genes directly dysregulated. This seems to be especially true in the case of the t(4;14) where two enzymes are overexpressed: FGFR3, a tyrosine kinase receptor, and MMSET, which has homology to histone methyltransferases. As a surface receptor, FGFR3 may be targeted by monoclonal antibodies, and as a tyrosine kinase, by selective tyrosine kinase inhibitors. Preclinical studies have validated FGFR3 as a therapeutic target in t(4;14) MM, and plans for a clinical trial are under way (S Trudel and L Bergsagel, unpublished). Histone methyltransferases are being developed, and studies are under way to validate MMSET as a target in MM.

Concluding Thoughts

The past year has seen the synthesis of a number of different observations into a clearer picture of the molecular pathogenesis of MM. There appear to be two pathways of pathogenesis that are associated with specific cytogenetic features, one that is hyperdiploid, and usually lacks primary Ig translocations (TC2 and perhaps some TC3), and one that is nonhyperdiploid and has primary Ig translocations (TC1, TC4, TC5, and perhaps some TC3). Subtypes are identified on the basis of the Ig translocation and cyclin D expression (TC classification). This dichotomy between hyperdiploidy and translocation superficially appears very similar to what is seen in all, and the analogy deserves further scrutiny.

The TC classification identifies clinically important molecular subtypes of MM with different prognoses and with unique responses to different treatments (e.g., high-dose therapy (HDT) and 11q13-TC1, FGFR3 inhibitor and 4p16-TC4). Prospective studies will be required to validate this classification for use in clinical trials. Almost all of the preclinical studies of new agents in MM are based on the induction of apoptosis at 24 or 48 hours in 1 or 2 cell lines (primarily 8226 and MM.1, both with 16q23 translocations TC5). It now is clear that there are a minimum of 5 different molecular subtypes of MM, and an effort should be made to include the more common ones in preclinical studies (e.g., at a minimum, U266 for 11q13-TC1, H929 for 4p16-TC4, and 8226 or MM.1 for maf-TC5). It also is obvious that we desperately need a model for the large group of patients (TC2) for which we do not appear to have representative cell lines.

For the reasons stated above, this may depend on developing models that are able to reproduce or replace critical features of the host BM microenvironment. Importantly, we need to gain a more comprehensive understanding of the molecular phenotype of each subtype of MM tumour, including identification of changes that remain essential for survival and growth of the tumour, thereby representing potential therapeutic targets. Although much remains to be learned, it seems essential that we rapidly incorporate what we have already learned about the molecular biology of myeloma into the clinical arena.

New Insights in the Biology and Treatment of Myeloma-Induced Bone Disease

MM is a disorder in which malignant plasma cells accumulate in the BM and usually produce a monoclonal immunoglobulin (Ig) of IgG or IgA type. MM is responsible for about 1% of all cancer-related deaths in Western countries and epidemiological studies have shown that at least one third of MM emerge from a preexisting benign monoclonal plasma cell disorder, i.e., MGUS. One prominent feature in MM is the occurrence of skeletal events including bone pain, pathological fractures secondary to lytic bone lesions, and hypercalcemia. Up to 80% of patients with MM present with bone pain, and over 70% of the patients will develop pathologic fractures during the course of their disease. Bone lesions and hypercalcemia correlate directly with the presence of the total mass of myeloma cells and have prognostic value. The excessive bone resorption is an early symptom in MM and a hallmark of malignancy in individuals with MGUS.

Bone Disease in MM

Myeloma cells grow in the BM, where the microenvironment supports their growth and protects them from apoptosis. The accumulation of myeloma cells within the BM is associated with increased rates of bone turnover. Histomorphometric analyses of bone biopsies from patients with MM have shown that an unbalanced bone remodelling formation was the characteristic feature of patients with osteolytic bone lesions, which on one hand increased osteoclastic resorption and on the other hand lowered bone formation.,

Increased Osteoclastic Resorption

A significant increase both in the recruitment of new osteoclasts and in single osteoclast activity occurs in the close vicinity of myeloma cells, suggesting that bone disease results from local production of an osteoclast activating factor (OAF) secreted by either myeloma cells or BM stromal cells. Recently, two kinds of factors have been identified as such OAF: RANKL/OPG system and the chemokine macrophage inflammatory protein-1 (MIP-1).

RANKL/OPG: Evidence from gene-deleted and transgenic mice indicates that generation of activated osteoclasts from monocytic precursors is controlled by coordinate expression of the RANKL (receptor activation of NF-êB ligand, also known as TRANCE, OPGL, TNFSF11) and its decoy receptor osteoprotegerin (OPG, TNFRSF11B). RANKL is expressed by osteoblastic cells and binds to its receptor (RANK) present on osteoclastic cells triggering differentiation and activation signals in osteoclast precursors, thereby promoting bone resorption. Osteoprotegerin (OPG) is a naturally occurring factor that antagonizes the effects of RANKL, thereby preserving bone integrity.

Therefore, the ratio between RANKL and OPG (RANKL/OPG) is determining to regulate osteoclast activity and bone resorption. Moreover, it has been suggested that the interactive network of bone-resorbing and antiresorptive cytokines and hormones converges at the RANKL/OPG system. RANKL/OPG then serves as the final common effector system to regulate osteoclast formation from precursors in the BM and its subsequent activation. Because the RANKL/OPG system is likely to play a pivotal role in the control of bone resorption, this axis was evaluated in MM-induced osteolysis. Recent studies have shown that myeloma cells are able to induce increased RANKL expression and decreased OPG production in the BM environment. First, an overexpression of RANKL has been observed in BM biopsies from patients with MM.

RANKL is overexpressed in stromal cells at the interface of MM with normal BM elements, rather than in myeloma cells. RANKL also may be produced by myeloma cells in some patients as described by Heider et al. Of note, in contrast to human myeloma cell lines, RANKL was detected in the murine myeloma cell line 5T2MM. In vitro coculture experiments have indicated that myeloma cells were able to induce RANKL expression in stromal/osteoblastic cells in part through cell-to-cell contact involving the integrin VLA-4 and in part through a soluble factor. Several OAF, including IL-1â, IL-6, and tumour necrosis factor (TNF)-a, have been reported as overproduced by stroma in response to MM. However, the RANKL overexpression seems unrelated to these cytokines since the addition of blocking antibodies against IL-1â, IL-6, and TNF-a in cocultures did not prevent RANKL upregulation.

The soluble RANKL-inducing factors involved in MM are still unidentified but may implicate IL-7, which is produced by myeloma cells. Such abnormalities of RANKL overexpression in bone environment participate in the pathogenesis of various osteolytic diseases especially osteolytic metastasis in breast cancer, in which the parathyroid hormone (PTH)-related peptide plays a major role. In addition to increased expression of RANKL, MM-infiltrated BM exhibit decreased production of the natural RANKL inhibitor OPG. Two mechanisms have been involved in that process. First, a decrease of OPG production by stromal cells has been described as induced by MM cells. Second, myeloma cells sequestrate OPG, internalise, and degrade this factor within the lysosomal compartment.

This process is dependent on physical interactions between OPG and heparane sulfates present on syndecan-1 highly expressed on myeloma cells. Both mechanisms may contribute to low local and systemic OPG levels observed in patients with MM. In summary, inhibition of OPG production at both transcriptional and post-transcriptional levels by myeloma cells associated with increased expression of RANKL in BM deeply disrupt RANKL/OPG ratio in favour of the osteoclastogenic factor RANKL. Finally, the main role of RANKL/OPG axis deregulation in MM-induced osteolysis is highlighted by the high potency of RANKL inhibitors such as OPG or RANK-Fc to prevent both excessive osteoclast development and lytic bone lesion occurrence in different murine myeloma models. Furthermore, as discussed below, disruption of RANKL/OPG axis may promote tumour progression, since treatment of mice with RANKL antagonists decreased tumour burden.

Chemokines MIP-1: Two different groups have recently shown that the chemokines MIP-1a and -â significantly participate in myeloma-induced bone disease. The first group found that MIP-1a was overproduced in myeloma BM and the second that both MIP-1a and MIP-1â were secreted by myeloma cells. The chemokines MIP-1 belong to the RANTES family and act as chemoattractants and activators of monocytes. Both osteoclast precursors and stromal cells express the chemokine receptor for MIP-1a and MIP-1â (CCR5). Data demonstrated that MIP-1a as well as MIP-1â induce expression of RANKL in stromal cells and consequently enhance osteoclast formation and resorbing activity. In line with these results, administration of neutralising anti-MIP-1a antibodies to 5TGM1 myeloma–bearing mice limited development of osteolytic lesions and intact RANK/RANKL signalling is necessary since MIP-1a had no effect in RANK null mutant mice. Choi et al recently cloned the human MIP-1a promotor and characterised the transcription factor (TF) that controls MIP-1a expression in MM cells.

They reported that the ratio of both alternatively spliced variants of the TF acute myeloid leukemia-1 (AML-1), AML-1A and AML-1B, regulates MIP-1a, and that abnormal expression of these TF in MM correlates with increased MIP-1a expression. The clinical correlation between severity of bone lesions and MIP-1 production by MM cells corroborates these results. In addition, the gene expression profile study of 92 primary MM indicated that the MIP-1a gene is overexpressed in osteolytic MM. Furthermore, because the chemokine receptor, CCR5, is also expressed by MM cells, MIP-1a and MIP-1â may act on MM cells in an autocrine paracrine fashion. In fact, it has been recently shown that MIP-1a triggers migration and signalling cascades mediating survival and proliferation in MM cells.

In addition to their osteoclast-inductive capacity, MIP-1a and MIP-1â have other biologic activities that may be relevant to clinical features of patients with MM. In fact, these chemokines have been suggested to be potent modulators of hematopoiesis: MIP-1a inhibited early erythropoiesis and MIP-1â increased apoptosis in pre-B cells. Therefore, MIP-1a and -â are pluripotent chemokines that may play important roles in the pathogenesis of several clinical features of MM including not only destructive bone lesions, but also suppression of erythropoiesis, of B lymphopoiesis and of immunoglobulin production.

Direct Interaction with Osteoclasts: Several studies have demonstrated that myeloma cells enhanced osteoclast formation and activity through osteoblastic cells (i.e. RANKL and MIP-1a). Moreover, studies on mice demonstrate the dependence of myeloma cells on osteoclast activity and highlight the importance of the myeloma-osteoclast loop for sustaining the disease process. But direct interactions between myeloma cells and osteoclasts remain unclear. Like this, the chemokine MIP-1a, in addition to acting through osteoblastic cells to enhance osteoclast activity, may also be a potent osteoclastogenic factor that acts directly on osteoclast precursors that express CCR5 to induce late stage differentiation. Furthermore, the gene coding for Gas6, the ligand for the receptor tyrosine kinase Tyro-3, is overexpressed in plasma cells. As Tyro-3 is expressed on mature osteoclasts and involved in stimulation of osteoclastic bone resorption, overproduction of its ligand by plasma cells may be at the origin of strong direct interactions between myeloma cells and osteoclasts. These observations suggest that an interdependence could truly exist between myeloma cells and osteoclasts but further data are needed to sustain this hypothesis.

Decreased Bone Formation

Histomorphometric studies and biochemical indicators of bone turnover in MM have shown that although osteoclast number and function are increased in MM, the key difference in vivo between the presence and absence of lytic lesions is that osteoblasts are fewer and less active in patients with lytic lesions. In the early stages of MM, bone formation is increased reflecting the coupling of resorption to formation. However, as the disease progresses, bone formation is decreased and this leads to an uncoupling resorption and formation and rapid bone loss.

This suggests that myeloma cells could first stimulate osteoblastic function during the early stages of the disease then inhibit it or even be toxic for these cells during overt expansion of the tumour. Few inhibiting interactions between osteoblasts and MM have been described so far. Recently, Shaughnessy et al reported the production of the potential osteoblast inhibitor DKK1 by myeloma cells. Actually, DKK1 can block Wnt signalling, an important pathway involved in osteoblast differentiation and function, and its overexpression in MM is associated with lytic bone disease.

However, further experiments need to be done to confirm these data. Other potential means for the interplay between osteoblasts and myeloma cells could be through homophilic binding by the neural cell adhesion molecule NCAM/CD56. On the one hand, NCAM is known to be overexpressed by MM cells mainly of kappa subtype, in correlation with the presence of lytic bone lesions. Conversely, the lack of or weak expression of NCAM by MM cells delineates a subset of MM at diagnosis mainly characterised by a lambda light chain subtype, a lower osteolytic potential and a trend for malignant cells to circulate in the peripheral blood. Of note, NCAM is also strongly expressed by human osteoblasts. Thereby, NCAM-NCAM homophilic binding between CD56⁺/NCAM positive MM cells and osteoblasts may induce a decrease in osteoblast function as we previously described for osteocalcin production. On the other hand, such negative interactions lacked in CD56⁻/NCAM negative MM.

Biochemical Markers of Bone Turnover

Classic biochemical markers of bone turnover remain poor predictive parameters in MM. Therefore, as RANKL/OPG axis and MIP-1a are particularly involved in the biology of bone resorption, their study as bone markers could be useful. First, median OPG serum levels were lower in patients with MM at the time of diagnosis than in healthy age- and sex-matched controls. Moreover, OPG levels were correlated with serum levels of carboxy-terminal propeptide of type I procollagen (PICP bone formation marker) but not with clinical stage or survival. Second, serum levels of sRANKL were elevated in patients with MM and correlated with bone disease.

The ratio sRANKL/OPG was also increased and correlated with markers of bone resorption (TRACP-5b, NTX), osteolytic lesions, and markers of disease activity (â2-microglobulin but not CRP). Furthermore, Abe et al found that MIP-1 production by MM cells in BM correlated with the severity of bone lesions. In conclusion, these markers may have a clinical utility.

New Therapeutical Approaches in Myeloma-Induced Osteolysis

The development of lytic bone lesions is a major cause of morbidity in patients with MM. However, the therapeutic arsenal available to control excessive bone resorption remains insufficient, despite the emergence of new bisphosphonates. The recent description of the RANKL/OPG system and its main role in bone remodelling regulation has opened new avenues in the therapeutic approach of excessive bone resorption.

In the close future, MIP-1 inhibitors could also represent a new effective therapeutical target to treat MM-induced bone disease. The 5T mouse model using the murine MM lines (5T2, 5T33, and its derived subclone 5TGM1) and the SCID-hu model have provided important tools for validation of in vitro observations and for preclinical studies. MM growth in the BM microenvironment was observed together with the appearance of osteolytic destruction of the human bone. The effect of potential antiresorptive drugs on bone disease and tumour growth has been evaluated in these models. These studies demonstrate that using bisphosphonates or blocking RANKL and the MIP-1a axis strongly affect not only bone resorption but also tumour development.

Bisphosphonates

Bisphosphonates, which are potent inhibitors of bone resorption, are widely used in MM-associated hypercalcemia. Placebo-controlled studies, generally including patients with stage III MM, have shown that bisphosphonates, mainly clodronate, pamidronate, and zoledronate, contribute to the long-term control of bone disease., They reduced the incidence of skeletal events, prevented hypercalcemia, alleviated bone pain, and improved the patient's quality of life. But they neither induced bone lesion healing nor improved the survival of patients, certainly because of their use at a too advanced stage of the disease Even though in vitro studies demonstrated that nitrogen-containing bisphosphonates induce apoptosis using human MM cell lines and in vivo use of pamidronate or zoledronate in the SCID-hu model and zoledronate in 5T2 halted MM bone resorption and decreased tumour burden, there is no proof so far that bisphosphonates really improve survival in vivo in patients.

Interestingly, it has been recently shown that both pamidronate and zoledronate stimulate OPG production by primary human osteoblasts. These observations strongly argue for the early use of bisphosphonates in MM to prevent bone disease and slow down tumour development.

RANKL Inhibitors

The treatment of mice with MM-associated bone disease (5T2 and SCID-hu) with recombinant OPG or RANK-Fc, a fusion protein of the extracellular domain of the murine RANK with the constant region of human resulted in a significant reduction in the number of osteolytic lesions. Histologic analysis demonstrated that OPG or RANK-Fc treatment was also associated with a partial preservation of cancellous bone volume and a significant increase of total bone marrow density.

In addition to inhibiting the development of MM bone disease, OPG or RANK-Fc treatment resulted in a decrease in serum paraprotein, raising the possibility that they may have an antitumour effect. This effect was not direct since neither OPG nor RANK-Fc could induce myeloma cell death by themselves but rather depended on decreased bone remodelling. Similar results obtained in SCID-hu hosts treated with bisphosphonates also support the notion that antimyelomatous effect of RANKL inhibitors or bisphosphonates is closely related to inhibition of osteoclast activity. RANKL inhibitors have been successfully used to treat osteolytic metastases, tumour-induced bone pain, and humoral hypercalcemia in various animal models of nonmyeloma malignancies. Furthermore, 2 studies have evaluated the skeletal effect of a single dose of first OPG in a series of postmenopausal women with increased bone turnover and second OPG construct (AMGN-0007) in MM patients with lytic bone lesions. In both cases, OPG caused a rapid, sustained dose-dependent decrease of bone resorption as indicated by a decrease of biological bone turnover markers.

The treatment was well tolerated and without adverse effects. However, bone mass, number of osteolytic lesions, and patient survival have not been assessed. Importantly, OPG inhibits TNF-related apoptosis-inducing ligand (TRAIL) induced apoptosis and can function in vitro as a paracrine factor for human MM cells. Altogether, these observations strongly suggest that RANKL inhibitors may represent a novel interesting approach to the treatment of MM-induced bone disease and various osteolytic diseases, but further studies should be done to evaluate adverse effects of OPG tumour protection against TRAIL-induced apoptosis.

MIP-1 Inhibitors

Because the chemokines MIP-1a and MIP-1â on the one hand are among the leading candidates for MM-derived factors that enhance osteoclast differentiation and function and on the other hand may also be involved in the development of major clinical features of MM such as anemia and hypogammaglobulinemia, inhibition of their production or activities could be a novel and powerful therapeutic target in MM. Such inhibitors could come from antibodies blocking MIP-1 chemokines or their receptors, i.e., small-molecule nonpeptide receptor (CCR5) antagonists or modified chemokines.

Conclusion

RANKL and OPG play an essential role in osteoclast formation and activation, and various bone tumours act through that system to trigger bone resorption. As described in this review and summarised in the interaction of MM with stroma results in deregulation of the RANKL/OPG axis, both in increasing RANKL and decreasing OPG. This disruption of the RANKL/OPG ratio in the bone environment increases osteoclast activity, triggers bone destruction, and promotes tumour growth. Moreover, the chemokines MIP-1a and MIP-1â produced by myeloma tumour also enhance osteoclast activity both through RANKL expression in bone environment and direct effect on osteoclast precursors leading to increased bone resorption. Finally, in vivo use of osteoclast inhibitors (bisphosphonates or specific inhibitors of RANKL) halted MM-induced bone resorption and resulted in inhibition of myeloma growth and survival.

These observations demonstrate a strong interdependence between myeloma cells and osteoclasts: myeloma cells enhance the formation of osteoclasts, whose activity or products, in turn, are essential for the survival and growth of myeloma cells. In line with this concept, a recent study has shown that IL-6 and osteopontin highly produced by osteoclasts played a central role in survival and growth of myeloma cells., Indeed, the use of effective osteoclast inhibitors in vivo could break down this vicious circle and both suppress bone resorption and decrease tumour growth.

It is tempting to speculate that interfering with BM cultivation by myeloma cells may inhibit the development of myeloma especially in early or premalignant stages (MGUS). Altogether, these observations strongly suggest that reducing the RANKL/OPG ratio by treatment with RANKL inhibitors and/or MIP inhibitors should provide a high therapeutic interest to decrease both bone resorption and tumour burden in MM.

Therapeutic Targets

Cancer cells arise through the progressive acquisition of mutations that deregulate cell cycle checkpoints, inactivate DNA repair mechanisms, disrupt apoptosis pathways, and alter host-tumour interactions allowing invasion and metastasis. Because of the acquisition of genetic lesions that perturb normal cell physiology, the hypothesis has been put forth that tumour cells differ from normal cells in their apoptotic burden. In normal cells, genetic abnormalities are "sensed," and signalling pathways are activated that lead to cell cycle arrest or the induction of apoptosis.

The "rheostat"' that regulates activity of the core cell death machinery (i.e., the threshold of stimuli required to induce apoptosis) is set at a low level to ensure that renegade cells with genetic lesions conferring a growth advantage are destroyed. As tumour cells evolve, they acquire (through the process of mutation and selection) a myriad of mechanisms enabling them to survive even in the face of death signals that should lead to their demise.

Because of this increased apoptotic burden, it can be sized that tumour cells live on the precipice of apoptosis: even partial inhibition of antiapoptotic mechanisms operative in tumour cells would be expected to render them vulnerable to cell death. Therefore, therapies designed to downregulate antiapoptotic pathways would be expected to enhance the demise of cancer cells without affecting normal cells. This hypothesis will be testable when therapeutic agents are developed that specifically target the key regulatory elements of the apoptotic machinery.

The Bcl-2 antisense oligonucleotide Genasense may represent the first of this class of agents, as discussed below. Note also that since most cytotoxic drugs exert their antitumour effects by activation of apoptosis pathways, interventions that inhibit antiapoptosis mechanisms in tumour cells should overcome many forms of drug resistance.

We believe that MM fits this paradigm. From the earliest stages of proliferation and differentiation in the lymph node germinal centre, incipient MM cells accumulate genetic lesions including translocations involving the Ig-H switch region, and gross chromosomal abnormalities leading to aneuploidy. Late in the disease, MM cells lose the requirement for the BM microenvironment as they acquire additional mutations involving oncogenes such as myc, ras, and p53. One would hypothesize that even during the earliest stages of disease (MGUS), genetic lesions result in an increased apoptotic burden, and the surviving cells must have developed antiapoptotic mechanisms to counterbalance the death signals.

This concept is consistent with the notion that while MM is a disease of deregulated proliferation, early in the disease the labeling index is low (< 1%), and an increased survival of malignant plasma cells may be a more important factor for the initial expansion of malignant plasma cells in the BM.

Apoptosis is induced through two distinct yet intertwined pathways: (1) the extrinsic or death receptor pathway composed of tumour necrosis factor (TNF)-family receptors and ligands and (2) the intrinsic pathway in which the release of mitochondrial constituents regulates caspase activation. Although the mitochondrial pathway of apoptosis is the main topic of this review, it should be noted that death receptors 4 and 5 (DR4, DR5) are expressed on MM cell lines and primary MM isolates, and can efficiently activate the extrinsic pathway after binding of the ligand TRAIL. This leads to the rapid activation of caspase 8 that can either directly activate effector caspases 3, 6, or 7 or cleave Bid, leading to amplification of the apoptotic signal by recruiting the mitochondrial pathway.

Regulation of TRAIL-induced death can also occur at the level of the BM microenvironment, as OPG released by osteoblasts and other stromal cells can act as a decoy receptor for TRAIL, thereby blocking its apoptosis-inducing activity. MM cells inhibit OPG release by stromal cells, thereby promoting osteoclast activation and lytic bone disease (by enhancing RANKL availability), while at the same time exposing themselves to higher levels of ambient TRAIL. As a recurring theme, the relative levels of proapoptotic versus antiapoptotic molecules that act in a cell autonomous manner or in the milieu of the BM microenvironment determine the outcome of potentially lethal signals.

Regulation of the Mitochondrial Pathway of Apoptosis

Myeloma cells are exposed to multiple noxious stimuli with the potential to induce apoptosis, such as chromosomal instability or hypoxia, and those induced by different forms of therapy, which work through mechanisms that are as varied as the therapeutic agents themselves (e.g., dexamethasone, melphalan, thalidomide, Velcade). Resistance to apoptosis in these cases ultimately rests on the ability of the MM cells to prevent activation of the mitochondrial pathway of apoptosis.

Whether noxious signals activate this pathway is determined by members of the Bcl-2 family, and understanding the molecular functions of these proteins is required for the design of novel therapeutics to overcome the resistance to apoptosis exhibited by MM cells.

Bcl-2 family members are divided into 3 functional groups; these encode 1 or more Bcl-2 homology domains (BH1-BH4) and act as inhibitors or inducers of the mitochondrial apoptosis pathway Antiapoptotic family members (Bcl-2, Bcl-XL, Mcl-1) are localised to the outer mitochondrial membrane (OMM) via a hydrophobic carboxy-terminal tail, and regulate the release of apoptotic molecules from the intermembrane space. The apoptosis inducers (e.g., Bax and Bak) encode BH1, BH2, and BH3 domains, and can be induced by apoptotic signals to homo-oligomerize and form pores in the outer membrane, thus permitting efflux of apoptosis-inducing molecules including cytochrome c, dATP, SMAC/Diablo, and AIF.

The mechanism by which mitochondrial outer membrane permeabilisation (MOMP) allows efflux of apoptogenic proteins is controversial; however, it has been shown that Bax oligomers can form pores in liposomes that permit the passage of cytochrome c. Under normal growth conditions, Bak is tethered to the mitochondrial outer membrane, while Bax translocates to the mitochondria in response to apoptosis-induced conformational changes that unmask its carboxy-terminal hydrophobic domain.

Cells isolated from Bax/Bak double knock-out mice exhibit a dramatic resistance to the induction of apoptosis by many different noxious insults, including overexpression of BH3-only proteins.

Domains of Bcl-2 Family Members

Transmembrane (TM) domains mediate insertion into the mitochondrial outer membrane. The Bcl-2-homology (BH) domains 1, 2, and 3 of antiapoptotic family members form a hydrophobic binding pocket for BH3-only proteins. Mcl-1 encodes a unique amino terminal domain of 180 amino acids, which may be involved in functions that are unique to Mcl-1.

The multidomain killers Bax and Bak can oligomerize to form pores in the outer mitochondrial membrane in a process regulated by BH3-only family members.

BH3-only proteins promote apoptosis by monitoring the status of cell "health" from different locations within the cell Thus, Bid is activated by caspase cleavage in response to signals from TNF-family death receptors, BAD senses the activity of growth factor regulated survival pathways (i.e., acting through Akt, PKA, and other kinases), Bmf and Bim appear to monitor microfilaments and microtubules, respectively, while Noxa and Puma are p53-regulated genes that respond to DNA damage.

The BH3 domains contain an amphipathic a-helix that functions as a "death domain" in these proteins, and is essential for both proapoptotic activity and, not coincidentally, the ability to bind to multidomain Bcl-2 family members. Whether a cell undergoes programmed cell death (PCD) in response to a potential apoptosis-inducing signal depends on the interactions of the BH3-only proteins with mitochondrial localised multidomain Bcl-2 family members. If the BH3-only proteins are sequestered by Bcl-2, Bcl-XL, and Mcl-1, apoptosis is prevented.

If specific BH3-only family members are able to interact with Bax or Bak, then oligomerisation is induced and cytochrome c is released.

The Induction of Apoptosis by Diverse Death Stimuli Occurs Through the Transcriptional or Post-transcriptional Regulation of BH3-only Proteins

Levels of Noxa and Puma are transcriptionally upregulated by p53 in response to DNA damage. Cytokines and growth factors act as survival factors by inhibiting the proapoptotic activities of Bad and Bim through changes in the state of protein phosphorylation. Loss of cell adherence to extracellular matrix can lead to a form of apoptosis termed "anoikis," and this is regulated by BH3 proteins that associate with actin filaments (Bmf) or microtubules (Bim). The caspase 8-mediated cleavage of Bid to form tBid acts as a link between the extrinsic and intrinsic pathways of apoptosis.

BH3-only proteins translocate to the mitochondria, where they can bind to antiapoptotic (Bcl-2, Bcl-XL, Mcl-1) or proapoptotic (Bax, Bak) Bcl-2 family members to regulate mitochondrial outer membrane permeability (MOMP) and the release of cytochrome c, apoptosis-inducing factor (AIF), and SMAC/Diablo. Bax is cytosolic until death signals induce a conformational change followed by its insertion into the outer mitochondrial membrane with formation of oligomers.

Recent studies demonstrate that the level of Ca^{+2} stores in the endoplasmic reticulum (ER) and subsequent uptake of Ca^{+2} by the mitochondria play an important role in determining the threshold for apoptosis. This cyclical ER-mitochondrial flux of Ca^{+2} is regulated in part by Bax, Bak, Bcl-2 and other family members that are localised to these organelles. Abbreviations: IMS, inter-membrane space; SMAC, second mitochondrial activator of caspases; PKA, protein kinase A

Much has been learned about how Bcl-2 and Bcl-XL interact with Bax, Bak, and BH3-only proteins. The core 3-dimensional (3D) protein structure is conserved between Bcl-XL and Bcl-2, and is composed of a globular bundle of 5 amphipathic a-helices surrounding 2 central hydrophobic a-helices.

Of great functional importance is the formation of a hydrophobic binding groove by the BH1, BH2, and BH3 domains that is required for binding to BH3 domains and for survival functions. In addition, the 3D structures of Bcl-2, Bcl-XL, and Bax resemble the pore-forming subunits of bacterial toxins and have weak channel-forming activity for small ions through lipid membranes. As described above, pore formation by Bax and Bak may be the critical event in MOMP.

Recent elegant experiments have led to a revised model for the regulation of the mitochondrial pathway of apoptosis by BH3-only proteins. A critical aspect of this model is based on the relative affinities of distinct BH3-only proteins for antiapoptotic versus proapoptotic Bcl-2 family members, and the relative abundance of each class of protein within the cell. Letai et al have determined the binding affinity of Bcl-2 for a variety of peptides encoding the functional alpha-helical regions of killer protein BH3 domains, and have determined which of these peptides can directly induce cytochrome c release from purified mitochondria. BH3-domain peptides from Bid and Bim directly induced MOMP and cytochrome c release through a process that required Bax or Bak. A BH3-domain peptide from BAD could not induce cytochrome release directly. However the BAD peptide did bind to Bcl-2 with high affinity and could displace the lower affinity Bid peptide. A model was proposed that some BH3 only proteins (e.g.,

Bid and Bim) directly target Bax and Bak and induce pore formation; apoptosis can be averted if BH3-only proteins are bound by antiapoptotic Bcl-2 family members. A second class of BH3 proteins, represented by BAD, cannot directly induce pore formation, but can occupy the hydrophobic binding pocket of Bcl-2 (and presumably Bcl-XL or Mcl-1) thus enabling subthreshold levels of Bid to target Bax or Bak.

Model for the Regulation of the Mitochondrial Pathway of Apoptosis by Bcl-2-family Members

In the absence of apoptotic stimuli, Bak molecules reside in the outer mitochondrial membrane but do not oligomerize to form pores; Mcl-1 may associate with Bak to block Bak self-association.

Apoptosis-inducing signals promote the movement of BH3-only proteins to the outer mitochondrial membrane (OMM), where some BH3-only family members (e.g., Bid, Bim) transiently associate with Bak or Bax to induce a conformational change resulting in formation of homo-oligomers and induction of apoptosis.

Antiapoptotic proteins such as Mcl-1 and Bcl-XL (or Bcl-2, not shown) may be overexpressed in multiple myeloma (MM) cells and sequester BH3-only proteins and prevent their interaction with Bak or Bax, thus overcoming cell death signals. Inhibition of expression of antiapoptotic proteins in MM cells could overcome this tumour survival mechanism.

Antiapoptotic effects of Mcl-1 and Bcl-XL can be overcome by excess BH3-only proteins.

Resistance to apoptosis may be reversed by the development of low molecular weight, cell-permeable drugs (D) that bind to the hydrophobic BH1-3 pocket of Mcl-1 and/or Bcl-XL, thus preventing the sequestration of BH3-only proteins. The latter are then free to interact with Bak and Bax with the subsequent formation of pores and induction of apoptosis. Efforts are in progress to identify novel compounds that specifically target Mcl-1 (D1) or Bcl-XL (D2).

The regulation of the mitochondrial pathway is even more complex, as the Ca^{+2} content of the endoplasmic reticulum (ER) has recently been shown to determine a cell's sensitivity to apoptosis. Apoptosis induced by some stimuli is associated with emptying of ER Ca^{+2} stores with a concomitant increase in mitochondrial Ca^{+2} pools. The steady-state level of Ca^{+2} in the ER and its release by apoptotic stimuli are regulated in part by Bcl-2 family members (e.g., Bcl-2, Bax, Bak) localised to ER membranes. The "mitochondrial pathway of apoptosis" may therefore be regulated by 2 hits: the Ca^{2+} flux from ER stores to the mitochondria, and the induction of mitochondrial outer membrane permeability by BH3-only proteins, both of which are regulated by multidomain Bcl-2 family members.

The Mitochondrial Pathway in MM

In MM, defects in programmed cell death pathways are frequently caused by imbalances in expression levels of the Bcl-2 family of proteins. It has become clear that every nucleated cell requires protection by at least one prosurvival Bcl-2 homologue, and that the abundance of these guardians regulates tissue homeostasis. MM cells conform to this model; they express Bcl-2, Bcl-XL, and Mcl-1, and both clinical and in vitro data suggest important roles for these proteins in maintaining MM cell survival and in clinical resistance to therapy. Although aberrant switch recombination leading to the activation of multiple translocation partners plays an important role in the pathogenesis of MM, these translocations do not include Bcl-2 family members.

Nevertheless, Bcl-2 is expressed in many (but not all) MM cell lines and primary clinical isolates. Bcl-XL is expressed in most cell lines and clinical isolates, and was detected more often at the time of patient relapse and correlated with resistance to chemotherapy. Mcl-1 appears to be expressed in virtually all MM cell lines and in all clinical isolates thus far examined by numerous groups., In a comparison of the expression of Mcl-1 in MM cells from 150 patients to that of plasma cells from 31 normal volunteers, a statistically significant increase in the myeloma cells was demonstrated (J. Shaughnessy et al, personal communication). Thus it appears that most MM cells express some level of Bcl-2, Bcl-XL, and Mcl-1, and it remains to be determined if these are purely overlapping in function, or also have distinct activities to promote tumour cell survival in the face of a myriad of apoptotic stimuli.

Experimental approaches to ablate expression of Mcl-1, Bcl-2, and Bcl-XL have begun to address this issue. A number of laboratories have correlated the induction of MM cell apoptosis with decreased expression of Mcl-1. Mcl-1 mRNA and protein have a short half-life, and inhibition of Mcl-1 synthesis led to the rapid induction of apoptosis of MM cells. The main apoptosis-inducing activity of the CDK-inhibitor flavoupiridol can be linked to inhibition of CDK9/cyclin T1 with subsequent inhibition of transcription elongation; Mcl-1 was shown to be a candidate target gene for the rapid induction of apoptosis by flavoupiridol.

The critical role for Mcl-1 as a survival factor in MM has been demonstrated in vitro using antisense oligonucleotides (ASO) to specifically inhibit Mcl-1 expression. Mcl-1 ASO led to the rapid induction of caspase activity and apoptosis (within 3 hours in some cases) when used as a single agent, and killing was potentiated by the addition of dexamethasone. In the latter study, ASO-mediated inhibition of Bcl-XL or Bcl-2 did not induce apoptosis as single agents, even though expression of the molecular targets was shown to be significantly reduced.

However, addition of dexamethasone to Bcl-2 ASO-treated cells did promote apoptosis in some cell lines. Others have demonstrated that in MM cell lines with a low level of Bcl-2 expression, apoptosis can be induced using the specific Bcl-2 ASO G3139, and that killing was potentiated by dexamethasone or taxol. A second study cultured purified primary MM cells with high concentrations of G3139 (10 iM, corresponding to 56 ig/mL; in clinical studies serum concentrations during the 7 day infusion were 3–7 ig/ml) and demonstrated a significant reduction of Bcl-2 RNA and protein in most patients. G3139 alone was not toxic to these cells, but it enhanced killing by doxorubicin and dexamethasone.

It remains to be clarified if inhibition of Bcl-XL can promote MM cell death as only 1 study has critically examined this question. No studies have systematically evaluated the role of Bfl-1/A1 in primary MM cell isolates.

A number of signal transduction pathways have been shown to regulate the expression of antiapoptotic Bcl-2 family members in MM cells, and the role of IL-6 has been closely scrutinised. STAT3 was recently shown to be constitutively activated in primary MM cells, and was shown to induce the upregulation of Bcl-XL in the U266 cell line. A number of groups have determined that Mcl-1 is upregulated by IL-6, perhaps through the activity of the STAT3 pathway. However the role of STAT3 was based on experiments using the tyrosine kinase inhibitor AG490, whose specificity and mechanism of action are not at all clear.

Our group demonstrated that approximately one third of MM cell lines and primary MM cells responded to IL-6 with the upregulation of Mcl-1, while in the other two thirds of cases, Mcl-1 was expressed at high levels in the absence of IL-6, without further increase upon IL-6 addition. An analysis of archival BM specimens indicated that while Mcl-1 expression was detected by in situ staining in all samples, phosphorylated STAT3 was only observed in 48%. This supports the notion that in some primary MM cells, Mcl-1 is expressed despite the absence of STAT3 activation.

Although IGF-1 promotes MM cell proliferation and survival, it does not appear alter expression levels of Bcl-2 family members. Overall, the data are consistent with our current understanding of MM as a genetically heterogeneous disease, and therefore it is highly likely that the molecular mechanisms regulating Mcl-1 expression will also be heterogeneous.

In summary, data indicate that Mcl-1 is expressed in all MM, and that this is required for the viability of myeloma cells. This may related to the important role of Mcl-1 during B cell terminal differentiation in the germinal centre. Its expression at the time of B-cell proliferation, class switch, and somatic hypermutation in the germinal centre may be required for survival of developing plasma cells, and its expression may then be maintained during the evolution of MM cells.

This is consistent with studies of the IL-6-mediated differentiation of Epstein-Barr virus (EBV)-transformed B cells to IgG expressing cells where Mcl-1 expression peaked as cells expressed high levels of IgG, and then declined in association with the induction of apoptosis. Bcl-2 and Bcl-XL are also expressed in most primary MM isolates, and it is likely that all 3 antiapoptotic Bcl-2 family proteins play important roles in preventing activation of the mitochondrial pathway. Future studies will be focused on determining how these molecules differ in their antiapoptosis functions, and whether individual cases of clinical resistance to treatment can be overcome by inhibition of one or more of these prosurvival factors.

Antiapoptosis Therapy in MM: Targeting the Mitochondrial Pathway

Understanding the intricacies of BH3 domain interactions will enhance efforts to develop novel antimyeloma agents that engage the apoptosis pathways in tumour cells. Cytotoxic insults that induce apoptosis through activation of BID-like BH3-only proteins act in a mechanistically different way from stimuli that promote apoptosis through the activation of BAD-like BH3-only proteins. From a therapeutic perspective, drugs that mimic the action of the Bad-like BH3 domain might be expected to sensitise tumour cells to a variety of apoptotic stimuli.

In the case of MM, it remains to be determined which Bcl-2 family interactions are most important in regulating MM cell survival or inducing apoptosis in response to different environmental signals. For instance, it will be important to understand which BH3-only proteins are activated by individual chemotherapeutic agents. Does melphalan induce 1 or a few different BH3-only proteins? Does p53 status make a difference (it would be expected to, given the DNA damage–p53–Noxa/Puma connection)? What about dexamethasone, thalidomide, proteasome inhibitors, and other agents that are just entering preclinical or early phase clinical trials?

Which specific BH3-only pathways do these novel compounds activate in MM cells? It may be possible to unravel the molecular details of how therapeutic agents affect critical regulatory interactions between Bcl-2 family members. Insights into these mechanisms of action would be very valuable for planning future clinical trials in which novel agents will be combined; agents that target different BH3-only family members might be expected to synergize for activation of the mitochondrial pathway of apoptosis. Novel compounds may activate individual BH3-only pathways at subthreshold levels for MOMP; however, if multiple BH3-only members can be activated, inhibition by Bcl-2, Bcl-XL, and Mcl-1 may be overcome, and dramatic apoptosis may ensue.

Clinical Trials of Bcl-2 Antisense Oligonucleotides in MM

Antisense oligodeoxynucleotides (ASO) have been shown to downregulate Bcl-2 expression and induce apoptosis in a number of human tumour models, including mouse xenograft models of melanoma and prostate cancers. In vitro experiments utilising ASO suffer from a number of pitfalls, including the requirement for special methods to introduce the ASO into cells that cannot be used in patients (e.g., electroporation, lipofectins, streptolysin-O), the use of concentrations of ASO that are not clinically relevant, and poor experimental designs that lack the minimal controls such as negative control ASO (i.e., those with scrambled or reverse sequence).

Nevertheless, preclinical data do indicate antitumour activity for the Bcl-2 ASO G3139 (Oblimersen Sodium, Genasense; 18-mer phosphorothioate ASO directed at codons 1–6 of Bcl-2) and this has been introduced into clinical trials with the goal of sensitising chemotherapy-resistant cancer cells to apoptosis by downregulating Bcl-2. In MM, a Phase III study comparing dexamethasone alone with the combination of dexamethasone plus G3139 has completed accrual and the data are pending. Two Phase II studies combining Genasense with active myeloma regimens for the treatment of relapsed/refractory patients have been described in abstract form. G3139 was administered as a continuous infusion (5–7 mg/kg/day) on days 1–7 repeated every 3 weeks, combined with dexamethasone (40 mg po days 4–7 of each week of G3139) and thalidomide (100–400 mg po per day).

After 3 cycles, responding patients went on to a maintenance phase with G3139 and pulse dexamethasone given every 5 weeks with continued daily thalidomide. Seventeen heavily pretreated patients have now been enrolled, and 15 patients have completed the induction phase. Responses include 2 CR, 1 near CR, and 5 PR (53% response rate). Some responses were in patients who had previously failed dexamethasone plus thalidomide. Laboratory correlates have thus far failed to demonstrate a significant decrease in the level of Bcl-2 protein expression in BM-derived MM cells; however, these studies are ongoing.

A second study added G3139 (7 day continuous infusion at 7 mg/kg/day) to the VAD regimen and 8 patients were reported, including 6 who were refractory to previous VAD chemotherapy. Four patients achieved PR with the G3139-VAD combination. Examination of Bcl-2 levels in peripheral blood MM cells using flow cytometry demonstrated a 18% and 17% reduction on days 4 and 7, respectively. For both Phase II studies, the treatment was well tolerated, with G3139 associated toxicities including mild thrombocytopenia, fatigue, and worsening of underlying renal insufficiency. Although these early results are encouraging, further analysis of the data from these ongoing trials will be necessary before conclusions can be drawn about the critical scientific issues they pose. These include whether G3139 actually decreases Bcl-2 protein expression in the MM cells of treated patients, and whether this renders tumour cells more susceptible to combined treatment with regimens that are already known to have high response rates in MM.

Future Drug Development Efforts Targeting the Mitochondrial Pathway of Apoptosis

We believe that Bcl-2, Bcl-XL, and Mcl-1 are valid therapeutic targets in MM, and through the use of structural biology and high-throughput technologies, a number of low-molecular-weight, cell-permeable compounds have been identified that bind to the BH1-3 hydrophobic pocket of Bcl-2 or Bcl-XL and promote apoptosis by blocking the association with BH3-only proteins. Two groups have taken the published nuclear magnetic resonance (NMR) structural data of Bcl-XL and used computer programs to model the 3D structure of Bcl-2.

If the 3D structure of a target protein is known or can be predicted from the structure of a homologous protein, computer screening of chemical databases can identify potential ligands that are predicted to bind to specific regions of the target. Using this technique, a virtual screen of chemical databases led to the identification of compound HA14-1 (MW 409) that was shown to bind to Bcl-2 with an IC_{50} of 9 iM using a competitive fluorescence polarisation assay in which the novel compound was tested for displacement of a fluorescein labelled Bak BH3 peptide from Bcl-2. HA14-1 induced apoptosis of > 90% of HL-60 cells (which express high levels of Bcl-2) when added to cultures at 50 iM for 4 hours.

A second group performed a computer screen of 207,000 compounds from the NCI 3D chemical database and identified 7 compounds that bound to Bcl-2 with IC_{50} values of 1.6 to 14.0 iM. One compound (#6) had significant growth inhibitory activity (IC_{50} 4 iM for HL60) and induced apoptosis of cells that expressed high levels of Bcl-2 (such as HL-60, MDA-231), but not low Bcl-2 expressers (T47D, MDA-453). This paradoxical result can be explained by the hypothesis that cells that express high levels of Bcl-2 are dependent on this protein to prevent activation of the mitochondrial pathway. Binding of compound #6 to Bcl-XL was shown by NMR analysis to cause peak shifts corresponding to amino acids located in the BH1-3 hydrophobic groove, thus confirming targeted binding of the compound. Using a high-throughput assay based on fluorescence polarisation, a screen of a 16,320 compound library yielded two new classes of inhibitors of Bcl-XL. NMR analysis demonstrated that these compounds bound to different positions within the Bcl-XL hydrophobic binding pocket and would be expected to compete for binding of BH3 domains from killer proteins, thus inhibiting the antiapoptotic effects of Bcl-XL. In fact these compounds did induce apoptosis of JK cells at 30–300 iM, and apoptosis was blocked by overexpression of Bcl-XL in these and in FL5.12 cells.

Another approach was based on the observation that drugs that inhibit mitochondrial functions such as electron transport often induce apoptosis and bypass Bcl-2 protective effects. Antimycin A, an inhibitor of complex III of the electron transport chain, was shown to induce apoptosis of Bcl-XL-overexpressing hepatocytes by binding to the hydrophobic groove of Bcl-XL. Methoxy-antimycin A, a synthetic derivative that no longer inhibited electron transport, maintained activity through its binding to Bcl-XL and inducing loss of mitochondrial transmembrane potential. It was speculated that in cells with high levels of Bcl-XL, antimycin A might activate the latent pore-forming activity of Bcl-XL, thus converting this antiapoptotic protein into a killer.

Drug discovery efforts to identify small-molecule inhibitors of antiapoptotic Bcl-2 family members are well underway. Once lead compounds are identified, the molecular basis for their interaction with target proteins will be evaluated using NMR or x-ray diffraction, and predictions can be made for structural modifications to increase affinity and specificity. Advances in medicinal chemistry and computer technologies, combined with development of novel high-throughput assays and methods for measuring the affinity of small molecules for target proteins, will permit a rapid, iterative approach for the optimisation of lead compounds. Compounds will move from the "wet laboratory" to the virtual world, and back, as each generation of compounds is tested.

Given the high degree of homology between the hydrophobic binding pockets of antiapoptotic Bcl-2 family members, this very detailed structural approach will be important for the development of inhibitors that specifically target Bcl-2, Bcl-XL, or Mcl-1. Specific inhibitors will be of great value in determining which antiapoptotic Bcl-2 family members are most important for the survival of MM cells, and whether the functions of antiapoptotic family members are overlapping or serve unique functions in specific tumour cells. The known genetic heterogeneity of MM suggests that responses to these agents will not be uniform, and that alone or in combination with other therapeutic agents, targeted inhibition of Bcl-2 family members should provide a profound reduction of the apoptotic threshold of tumour cells leading to improved therapeutic outcomes.

The possibility of inducing apoptosis of normal cells will require that great care be taken in the initial evaluation of these agents. Drugs that simultaneously target all 3 antiapoptotic proteins, while lethal to tumour cells, are likely to be very toxic to the host; specific agents are therefore required.

Improving Disease Control in Myeloma

This review of recent progress in the treatment of myeloma addresses high-dose therapy with autologous stem cell support, allogeneic transplants, new drugs, and their roles in disease management. Special scenarios of advanced age, renal failure, and 1° amyloidosis will also be covered. An algorithm for disease management based on prognostic factors will be presented.

High Host Risk Disease

The dose-limiting toxicity of stem cell–supported high-dose melphalan is stomatitis, which causes greater morbidity and mortality in the setting of renal failure, advanced age, and primary amyloidosis. Melphalan dose reduction to 140 mg/m^2 or even 100 mg/m^2 virtually eliminates this problem so that such patients are now frequently offered a dose-intensive treatment approach as well. Boccadoro and colleagues had demonstrated superior outcome with MEL 100 in 71 patients aged 60 to 70 years in comparison to a historical standard melphalan-prednisone control. These results have since been confirmed in a prospective randomised trial. In a comparison of 2 cycles of MEL 100 versus historically controlled pairmates receiving a two (single) cycles of MEL 200, the authors demonstrated comparable OS albeit inferior EFS with the dose-divided regimen.

The setting of primary amyloidosis, with or without myeloma, is particularly challenging. Recent data from Mayo Clinic, also presented at Salamanca, comprised 125 patients with primary AL involving kidney (66%), heart (48%), peripheral nerves (14%), and liver (17%). The majority received G-CSF for PBSC collection. Conditioning was mainly with MEL 200 in 59, MEL 140 in 29, and MEL 100 in 11 patients; the remaining 17 patients received MEL 140 + TBI. Reduced dose conditioning was done according to patient risk factors. Hematologic responses (defined as in myeloma) and organ responses (defined as functional improvement) both were higher with MEL 200 and MEL 140 + TBI regimens (74% vs 48%, $P = .01$). Patient outcome was dominantly affected by serum creatinine at time of transplant and the number of organs involved.

Median OS has been reached at 17 months for those patients with more than 2 organs involved by primary AL; organ responses were seen in 64% of patients, a result deemed superior to standard treatment approaches. Dose adjustments allowed for more patients with high-risk primary AL to be treated resulting, however, in a lower response rate. We recommend multiple cycles of high-dose therapy so that treatment can be tailored according to host risk and disease response, as it is now being evaluated in SWOG (S0115): following an initial cycle of MEL 100, the second cycle can be escalated to MEL 140 or MEL 200, in case the first cycle was well-tolerated and an insufficient antitumour effect was noted.

Issues and Non-issues with Autotransplants

Given the current state of the art, CD34 selection does not appear warranted as the level of tumour cytoreduction even with tandem transplants appears insufficient in the majority of patients for this procedure to impact outcome. Moreover, post-transplant immunosuppression appears prolonged, especially in the setting of CD 34^+, Thy^+, LIN^- superselection. Improving preparative regimens beyond single agent MEL 200 has been difficult. Any reduction in melphalan dose, required to accommodate additional agents (TBI, busulfan, carmustine, cyclophosphamide, etoposide), led at best to equal and often inferior results.

This is not to say that additional post- or peritransplant noncytotoxic approaches, such as dexamethasone, thalidomide or Velcade, might not improve treatment results by interfering with DNA damage repair or suppressing myeloma survival stimuli that originate in the bone marrow microenvironment. Such peri and intertransplant therapies are now being explored as part of IFM 99 and at Arkansas.

The value of intensive remission induction and post-transplant consolidation is currently under investigation in the Arkansas Total Therapy II (TT II) program that also examines the role of thalidomide in a prospective randomised trial design.

Results of the first 231 of currently 550 patients enrolled on TT II (accrual goal, 660 patients) were compared with the outcome of all 231 patients previously treated according to TT I. Data are still blinded with regard to thalidomide randomisation and are therefore presented for all patients.

Allogeneic Transplants

Standard myeloablative allogeneic transplants have been associated with high mortality often in excess of 50%. European Bone Marrow Transplant (EBMT) registry data, however, indicate marked improvement over the past 5 years so that survival has increased markedly, exclusively due to decreased transplant-related mortality with no reduction in relapse rate. Good prognosis was associated with female gender, younger age, low B2M, and responsiveness to prior therapy. An intergroup trial led by Southwest Oncology Group (S9321) evaluated, in patients up to 55 with an HLA-compatible sibling donor, a primary allogeneic transplant using MEL 140 plus TBI 12 Gy after VAD induction.

The study was closed due to excessive transplant-related mortality after enrolling only 38 patients. Remarkably, however, EFS and OS have stabilised beyond 5 years, whereas there have been continued relapses and disease-related deaths among patients registered on autotransplant and standard treatment arms.

Nonmyeloablative regimens (mini-allotransplants) permit complete donor hematopoietic engraftment, especially when given after prior melphalan based autotransplants. Mortality at 100 days has declined markedly to 10%–15% although, eventually, up to 25% to 30% succumb mainly due to complications from chronic graft-versus-host disease. Use of MEL 100 alone or in combination with fludarabine or TBI 200 cGy plus fludarabine, all resulted in only minor acute treatment-related toxicity and afforded prompt hematopoietic engraftment so that a third of patients did not require platelet or red blood cell transfusions. Similar positive results have also been reported for reduced intensity regimens in the context of matched unrelated donor (MUD) transplants.

Updates of results from Seattle, MD Anderson, and Arkansas revealed superior outcome when mini-allotransplants were conducted as part of a planned tandem transplant strategy, i.e., following a consolidative autotransplant with MEL 200. Disease recurrence, however, still occurs not infrequently, and it is too early to determine whether such mini-allotransplants, despite their lower acute toxicity, provide sustained disease control previously noted with standard allotransplants.

Prognostic Factors

Many prognostic factors, pertinent to standard dose therapy, have also influenced outcome with high-dose therapies. They include B2M, CRP, plasma cell labelling index, LDH, and cytogenetics. Metaphase cytogenetic abnormalities (CA) are usually informative in no more than one third of newly diagnosed patients, although all myeloma cells are aneuploid when examined by interphase molecular analysis. A direct comparison of interphase FISH and conventional cytogenetics as part of TT II revealed that, among those exhibiting FISH-based del 13, only patients with del 13 by conventional cytogenetics had inferior outcome; those with FISH-13 without CA had a survival similar to those without FISH 13; patients with other CA (no FISH 13, CA.

We recommend that MM with CA of any type be considered a separate “malignant myeloma” entity with a short 3 to 4 year median survival even with tandem transplants that deserve novel treatment concepts, including the tandem auto/mini-allotransplant approach.

Application of gene expression profiling to the majority of TT II patients at diagnosis has revealed unique expression profiles which can distinguish CA versus no CA and, within each subgroup, del 13 versus no del 13. Work is in progress to determine whether GEP of bone marrow biopsies together with analyses of highly purified CD138-positive plasma cells can help elucidate whether, in the presence of CA, the bone marrow microenvironment responds with amplified survival and proliferation signals far beyond levels evoked by more benign disease.

Previously Treated Myeloma

Such patients’ initial presentation, treatment details, and response require careful scrutiny with special emphasis on obtaining informative cytogenetics, since survival from relapse after tandem autotransplant is much shorter in case CA is present at relapse In case of pancytopenia, the possibility of t-MDS has to be considered, especially in patients treated for more than 1 year with melphalan or other stem cell toxic agents (e.g. carmustine). If cryopreserved PBSC are not available (especially when not previously autotransplanted but also when post-transplant survival exceeds 4–5 years), PBSC collection should be considered. Toward that goal, stem cell–toxic therapy should be avoided and DEX pulsing considered, possibly with the addition of low doses of thalidomide (50–100 mg).

Thalidomide

Thalidomide effects responses (e” 50% myeloma protein reduction [PR]) in one third of patients with advanced and refractory disease, the majority of whom had received prior autotransplants. Revimid is another immunomodulatory agent that exhibits virtually no sedative and only occasionally neurotoxic side effects. Responses have been reported in one third of patients with advanced and refractory myeloma. Many of these patients had been previously exposed to thalidomide although true thalidomide resistance was infrequently established.

Unlike thalidomide, Revimid causes myelosuppression which, in the setting of compromised bone marrow reserve due to extensive prior cytotoxic drug exposure, may not be fully reversible. In a Phase III trial for advanced myeloma comparing 2 different schedules of administration (50 mg × 10 doses and 25 mg × 20 doses q 28 days) we observed higher response rates with the more prolonged 25 mg dose schedule Grade > 2 thrombocytopenia was linked to pretreatment platelet count < 100,000/iL as a reflection of impaired of hematopoietic reserve a randomised trial is being initiated in SWOG (S0232) for nontransplant candidates, evaluating Revimid alone versus its combination with dexamethasone.

Velcade

The proteasome inhibitor, Velcade, effected PRs in one third of refractory myeloma patients, most of whom had been exposed previously to thalidomide and many had prior autotransplants. Velcade was combined with thalidomide (+ DEX) (VTD regimen) for posttandem transplant relapsed and refractory disease. The majority of patients had CA and had been exposed and become refractory to prior thalidomide VTD affected PR in approximately 60% of such patients regardless of the presence of CA (data not shown). Durable disease control, however, was limited to patients exhibiting no CA prior to VTD.

Gene Expression Profiling

Studies of gene expression profiling prior to and 48 hours after treatment have been helpful in distinguishing drug-specific and response-specific alterations in gene expression Importantly, response to Velcade plus thalidomide could be predicted with high accuracy.

The Arkansas Approach to Therapy

Subjects with MGUS and smoldering multiple myeloma are studied with DNA microarray and immunologically as part of a SWOG trial S0120. Those with smoldering multiple myeloma at high risk for progression to overt disease (single bone lesion, IgA isotype, higher M protein levels) are offered thalidomide plus dexamethasone plus bisphosphonates to determine whether such progression can be delayed/ prevented (S0231). This trial was developed based on encouraging data with thalidomide plus bisphosphonates observed in nearly 80 patients treated at Arkansas; similar results were noted at MD Anderson and Mayo Clinic.

Toward improving therapy of overt myeloma, a major focus should be on the high-risk entity presenting with CA. We envision the evaluation of Velcade, together with thalidomide + DEX (VTD regimen) in high-risk myeloma up-front, both as induction prior to and for maintenance after melphalan based tandem autotransplant. Assuming that the bone marrow microenvironment provides critical survival signals that rescue residual myeloma cells during hematopoietic recovery, peritransplant administration of DEX, thalidomide (THAL), or DEX + THAL, eventually together with Velcade, may provide an important adjunct toward extending disease control.

Good-risk patients may enjoy a median EFS exceeding 6 to 7 years when treated according to TT II. Given the remarkably high response rate to THAL + DEX for induction, reported recently by Mayo Clinic and MD Anderson investigators, this approach is currently being tested for induction and maintenance in the context of a standard tandem autotransplant with MEL 200 mg/m^2 under the auspices of SWOG (S0115). It is anticipated that such an approach is less toxic than chemotherapy induction and hence may increase the tandem transplant compliance, resulting in disease control comparable to more intensive approaches practiced with TT II.

Perspective and Conclusion

Minor permutations of standard melphalan-prednisone have long hindered advances in myeloma therapy. Raising the incidence of CR from previously less than 5% to the 50% level by melphalan dose intensification implies profound tumour cytoreduction well beyond the clinical detection level of 10^9 tumour cells in a sizeable fraction of patients. Intense investigations of patients with cytogenetic abnormalities, now also readily assessable by GEP, should help identify which of the new agents, targeting both myeloma and the bone marrow stroma, are most promising and can be incorporated into the MEL 200 based backbone of tandem transplants. Advances in such high-risk disease should provide important clues toward further refinement of therapy also for good-risk patients.

Molecular Diagnostics

It is increasingly evident that molecular diagnostics, that is, the use of diagnostic testing to understand the molecular mechanisms of an individual patient's disease, will be pivotal in the delivery of safe and effective therapy for many diseases in the future. A huge body of new information on the genetic, genomic and proteomic profiles of different hematopoietic diseases is accumulating. This chapter focuses on new technologies and advancements in understanding the molecular basis of hematologic disorders, providing an overview of new information and its significance to patient care.

In Section I, Dr. Braziel discusses the impact of new genetic information and research technologies on the actual practice of diagnostic molecular hematopathology. Recent and projected changes in methodologies and analytical strategies used by clinical molecular diagnostics laboratories for the evaluation of hematologic disorders will be discussed, and some of the challenges to clinical implementation of new molecular information and techniques will be highlighted.

In Section II, Dr. Shipp provides an update on current scientific knowledge in the genomic profiling of malignant lymphomas, and describes some of the technical aspects of gene expression profiling. Analysis methods and the actual and potential clinical and therapeutic applications of information obtained from genomic profiling of malignant lymphomas are discussed.

In Section III, Dr. Liotta presents an update on proteomic analysis, a new and very active area of research in hematopoietic malignancies. He describes new technologies for rapid identification of different important proteins and protein networks, and the potential therapeutic and prognostic value of the elucidation of these proteins and protein pathways in the clinical care of patients with malignant lymphomas.

Molecular Diagnosis of Hematopoietic Disorders

The monumental research advances in genomic and protein research over the past several years have made it possible to envision, in the not too distant future, the development of medical care that is truly tailored to each individual patient. Clinicians are anxious to incorporate this new knowledge into the selection of more specifically targeted therapies, and there is general agreement that new research insights need to be translated into useful clinical tests. Ideally, information from all available studies, including traditional morphologic and immunophenotypic findings as well as data from new genomic, proteomic, and pharmacogenomic studies would be applied to define a patient's disease profile.

This profile could be used not only to confirm and refine primary diagnoses, but potentially might also be used to determine the risk for development of cancer; to detect cancer at an early, more curable stage; to predict drug efficacy against a cancer and also drug toxicity; to improve disease staging for determination of risk of distant spread and subsequent relapse; and to assess response to therapy by monitoring for residual disease.

How close are we in 2003 to providing our patients with this type of individualised medical approach? Unfortunately, it must be acknowledged that we are still far from this ideal. Thus far, few of the exciting scientific advances of the recent past have been translated into improved patient care or better technology for molecular diagnostics. Nonetheless, the field of molecular diagnostics is growing, adapting, changing constantly, and beginning to permeate virtually every area of medicine provides a listing of some of the current and anticipated techniques used in molecular diagnostics today. These techniques are discussed below.

Tests for Genome-Wide Screening for Chromosomal Abnormalities

Routine cytogenetics is the traditional method for a survey of genome-wide chromosomal abnormalities, but standard karyotyping studies, even with chromosome banding, miss many subtle chromosomal abnormalities. Other methodologies for genomic profiling of chromosomal abnormalities have been developed, which have considerably augmented our knowledge of the genetic features of various hematopoietic malignancies.

These assays for genomic profiling are based on screening of chromosomes or DNA for loss or gain of chromosomes or genes, in contrast to gene-*expression* profiling performed on RNA. These assays detect changes in chromosome/gene location and number, not gene expression and function; for all of these studies, at least for disomic loci, the normal reference copy number is 2. Spectral karyotyping (SKY) and comparative genomic hybridisation (CGH) are complementary fluorescent molecular genetic techniques for detection of whole genome chromosomal abnormalities. With SKY, 24 differentially labelled painting probes representing all chromosomes are cohybridised, Fourier spectroscopy is used to distinguish the different spectrally overlapping probes, and special imaging software is used for analysis. This technique has been found to greatly facilitate the detection of many previously cryptic chromosomal translocations and rearrangements, and is already available for clinical purposes in many institutions. CGH uses the hybridisation of differentially labelled tumour DNA and reference DNA to produce a map of the DNA copy number changes in the tumour genome.

CGH assays are not yet available for routine clinical use, but technical permutations of this research methodology are reputedly in the pipeline for clinical laboratories. A variant of CGH, called matrix CGH, uses genomic cDNA fragments instead of the chromosome targets used in standard CGH, and even more powerful is the use of arrayed cDNA sequences with CGH. These latter techniques allow detection of unknown amplified genes, not just gene regions, and provide even higher resolution for identification of genomic imbalances.

Tests Targeting Specific Chromosomal Abnormalities

Multiple methods can be used for the detection of specific chromosomal abnormalities, including various permutations of the polymerase chain reaction (PCR), Southern blotting, and fluorescence in situ hybridisation (FISH) with molecular probes. PCR-based methods, often multiplexed, have been the screening test of choice for most molecular laboratories if the chromosomal abnormality of interest was amenable to PCR analysis. However, as the number of genes important in diagnosis and prediction of prognosis has increased almost exponentially over the past few years, a different molecular testing algorithm has evolved for hematopoietic malignancies.

The development, validation, and maintenance of numerous PCR analyses for detection of the ever-increasing important chromosomal abnormalities in hematopoietic malignancies is simply not practical for most laboratories. Fortunately, the development of molecular probes for use in FISH assays has provided a valuable alternative method to standard PCR analyses. FISH assays are not as sensitive as PCR assays, but FISH analyses are used predominantly at diagnosis and relapse, a time when only a low level of analytical sensitivity is needed since there are usually high levels of abnormal cells. The use of FISH assays for molecular evaluation of malignant lymphomas and leukemias has increased remarkably over the past 1–2 years, and has blurred the lines between classical cytogenetics and molecular pathology.

FISH is a very useful technique for detection of targeted chromosomal abnormalities. It can be done on blood, bone marrow, tissue touch preparations, body fluids, and even paraffin-embedded fixed tissue, so it is applicable to many specimen types. FISH overcomes one of the biggest problems with routine cytogenetic analysis of many lymphoma and chronic leukemia samples (i.e. the need for metaphases), as FISH can be done with either metaphase or interphase preparations. In FISH assays, the target is usually nuclear DNA of interphase or metaphase cells attached to glass microscope slides.

Most FISH assays are based on the ability of single stranded DNA to bind (hybridise) to complementary DNA, although some RNA FISH assays are available. The molecular test probes (DNA) can be labelled with biotin or digoxigenin labelled nucleotides and detected with fluorophor conjugated antibodies, or may be directly fluorophor labelled. With the use of dual or triple pass filters, multicolour FISH can be done.

There are several different strategies for the design of FISH assays. Single fusion-dual colour FISH assays for translocations utilise 2 probe hybridisation targets located on 1 side of each of the 2 genetic breakpoints; the usual level of false positive background cells from incidental overlap of signal in this type of assay is 5%–10%. Dual fusion-dual colour FISH assays for translocation utilise large probes that span 2 breakpoints on the different chromosomes. Dual fusion-dual colour FISH is optimal for detection of low levels of nuclei possessing a simple balanced translocation, as it greatly reduces the number of normal background nuclei with an abnormal signal pattern.

FISH using dual colour-break apart probes is very useful in the evaluation of genes known to have multiple translocation partners; the differently coloured probes hybridise to targets on opposite sides of the breakpoint in the known gene. Multicolour FISH using 3 to 4 differently coloured probes can be done in selected cases to determine the overlap of different genetic abnormalities in different cell populations. FISH with centromeric probes is useful for detection of changes in chromosome number (i.e., monosomy, diploidy, trisomy).,

Genomic probes for the genetic abnormalities of many leukemias, lymphomas, and even myeloproliferative and myelodysplastic disorders are now readily available from commercial sources. The most notable expansion in the library of molecular FISH probes is for the B-cell malignancies; some of the B-cell lymphoma-associated chromosomal abnormalities for which FISH analysis can be performed are listed in FISH assays are particularly useful in detection of chromosomal translocations that are not amenable to PCR detection due to widely distributed breakpoints, because FISH probes are much larger than the probes and primers used in PCR analysis. Like SKY, FISH assays will detect some genetic abnormalities that are karyotypically silent.

Non-random Chromosomal Abnormalities in B-cell Lymphomas

It should be remembered that FISH assays are useful mainly around the time of initial diagnosis or at relapse, when there is a relatively high level of abnormal cells. FISH is not useful for detection of low level minimal residual disease (MRD) following therapy, as the sensitivity of even the best dual fusion-dual colour FISH assay is only approximately 1 positive cell in 100 normal cells, not sufficient for detection of MRD. It is also important to remember that, despite the glamour of some of the newer techniques, standard cytogenetics continues to be extremely important in the initial diagnosis and follow-up of patients with hematopoietic malignancies.

Focusing only on tests that target specific genetic abnormalities, like FISH and PCR, can result in the failure to detect the additional important cytogenetic abnormalities that may be present initially or that may occur following therapy. For example, the need for intermittent cytogenetic analysis is very clear in chronic myelogenous leukemia (CML) patients. A number of these CML patients have developed clonal karyotypic abnormalities in Philadelphia chromosome–negative cells while on therapy with imatinib mesylate; these abnormalities would not have been detected by FISH or PCR analyses for BCR/ABL.

Genotyping for single nucleotide polymorphisms (SNPs) is relevant to both research and routine molecular diagnosis. Allele-specific PCR amplification techniques using sequence-specific primers (PCR-SSP) are widely employed for detection of SNPs in the genes that encode immunogenic proteins such as alloantigens. The status of certain alloantigens (i.e., human leukocyte antigens, blood group antigens, human platelet alloantigens) is frequently investigated by this methodology before organ transplantation. Technical advances in genotyping of SNPs have improved the ability to perform testing for HLA and blood group antigens on small samples of DNA, such as those obtained from patients with low leukocyte counts.

Gene Expression Profiling

Despite glowing predictions of future clinical utility and multiple published reports on the use of microarrays for gene expression profiling (GEP) in lymphomas and leukemias,, no microarrays are available yet for molecular diagnostics. However, focused arrays utilising fewer, but highly significant genes, are currently available for research studies and are in the development pipeline for B-cell non-Hodgkin lymphomas and acute leukemias. The diffuse large B-cell lymphomas are likely to be the lymphoma subtype for which focused microarrays will first be used for routine clinical purposes. Needless to say, molecular laboratories are anxiously awaiting the shift of this technology into the clinical arena.

Although microarrays for GEP have not yet made it into the clinical molecular laboratory, information gained from GEP data is impacting clinical laboratory testing. One example is flow cytometric analyses of *ZAP-70* expression in chronic lymphocytic leukemias/ small lymphocytic lymphomas (CLL/SLL). Pilot GEP studies in patients with CLL/SLL identified genes that were differentially expressed between leukemic clones that did not have mutated IgH_V regions and those that did. The best discriminator was a gene called *ZAP-70*; a high level of *ZAP-70* expression correctly predicted unmutated IgHv gene status in most patients.

This is clinically relevant because the absence of somatic mutations in the variable regions of the IgH gene has been determined to have adverse prognostic significance in CLL/SLL patients; those with unmutated IgHv regions often have progressive disease while those with mutated IgHv regions often pursue a more indolent course. Since molecular testing for IgHv mutation status involves multiple complex PCR reactions and sequencing procedures, the analysis is impractical for clinical testing. Fortunately, the ZAP-70 protein is readily detected by either flow cytometric analysis or immunohistochemical staining, and these procedures are currently being set up in many clinical laboratories in lieu of molecular analysis for somatic mutation of the IgH_V regions or microarray GEP.

There are some caveats about GEP data; these may be part of the reason for delay in implementation of this technique in clinical practice. Numerous GEP databases are now available in the public domain for different lymphomas, acute leukemias, and myelomas, and multiple publications have resulted from independent analyses of these databases. It turns out that different investigators do not always find the same results and draw the same conclusions from analyses of the same databases. This has made it apparent that, because of the extreme complexity, there are potential problems with analyses of GEP microarrays and databases that can produce erroneous GEP results. A few of the problems that have been described include sampling variability of tumours, chip differences and defects, differences and biases in analysis of GEP data, and sources of systematic error in microarray analysis. Clearly, sifting the real GEP changes from artifacts and noise in microarray experiments is often difficult, but rapid identification and neutralisation of spurious results is essential to prevent them from becoming accepted facts.

Another potential problem with GEP is the possibility of missing relevant cell populations present at a low level in the tumour specimen. Since GEP provides an average expression profile for an entire cellular population, small subpopulations of important cells are unlikely to be recognised. The application of new methods for microdissection, followed by RNA amplification, would allow targeting of specific populations of interest that were previously missed by GEP. In conclusion, there are enough problems with GEP microarrays and interpretations that it is important to have validation of significant GEP changes from more than one laboratory/database before important clinical decisions are based on this data.

Molecular Tests for Minimal Residual Disease Detection

Although many patients with hematologic malignancies achieve a complete clinical remission (CR) and even a complete pathologic remission by standard morphologic and immunologic criteria, a relatively high proportion of them will ultimately relapse. The source of this relapse is clearly from a persistent malignant cellular population that is present at a low level, below the limit of detection by standard techniques. For this reason, considerable effort has been devoted by molecular laboratories in the past 5 to 10 years to develop new molecular techniques to increase the sensitivity of detection of neoplastic cells.

The application of these techniques has demonstrated the presence of residual neoplastic cells in many patients in CR. This reservoir of neoplastic cells, detected only by sensitive molecular methods, is commonly referred to as minimal residual disease (MRD). The detection of MRD in a variety of hematologic malignancies suggests that obtaining a molecular remission should be a goal of therapy, and the results of most studies of MRD detection support this concept. However, it has still not been clearly established for many hematologic malignancies that patients with only a few residual malignant cells, detected only by very sensitive techniques, will benefit from additional therapy.

If achieving a molecular remission is confirmed to be an important goal following therapy for most hematologic malignancies, as seems likely, then MRD testing will become a much larger component of testing in molecular diagnostics laboratories. Ideally, techniques used for MRD detection should have a sensitivity level in the 10^5 to 10^6 range, be applicable to almost all patients with the disease, provide some quantification of the target, and be rapid, inexpensive, readily standardised, and disease-specific.

Also of critical importance for the clinical utility of tests for MRD detection is good interlaboratory reproducibility and standardisation of reporting. In reality, most commonly used molecular analyses for MRD detection do not meet many of these criteria. A particular problem for clinicians is the lack of standardisation of testing techniques and primers between laboratories, which essentially mandates follow-up testing for MRD be performed in the laboratory that did the previous testing to allow comparison of results. With frequent shifts in patient locations and changing insurance carrier requirements, sending follow-up specimens to the same laboratory may be impossible.

Only a few commonly used techniques are sensitive enough for detection of MRD in leukemias and lymphomas. Nested PCR and quantitative real-time PCR can be used for disease-associated translocations, without the need for patient-specific primers.

If the malignant clone does not carry a good translocation target for PCR analysis, patient-specific gene rearrangements may be targeted, using either nested or quantitative real-time PCR. Nested PCR analyses can detect up to 1 malignant cell in 10^6 normal cells. Quantitative real-time PCR assays, with a sensitivity of 1 in 10^4–10^5, are almost as sensitive as the nested PCR. A substantive number of studies of MRD detection have been performed in only a few hematopoietic malignancies, specifically chronic myelogenous leukemia, follicular lymphoma, and childhood acute lymphoblastic leukemia. The different methods used for detection of MRD in these 3 different hematopoietic malignancies are discussed below.

Chronic Myelogenous Leukemias

With imatinib mesylate therapy, a complete cytogenetic response (CCR) can be achieved for most patients with newly diagnosed chronic myelogenous leukemia (CML). Quantitative real-time RT-PCR analysis (Q-RT-PCR) is most often used to monitor for MRD in patients who have achieved a CCR by bone marrow cytogenetics and/or FISH. Interestingly, Q-RT-PCR monitoring for BCR/ABL can be performed on either peripheral blood or bone marrow; comparable results have been found on analysis of simultaneous blood and marrow specimens (R Braziel, unpublished data). This facilitates follow up of imatinib-treated CML patients.

Real-time PCR is a relatively new molecular technique that allows simultaneous PCR amplification and detection of target DNA or cDNA sequences. The specimen is normalised against an internal control, typically a single copy gene; for CML MRD testing, ABL or G6PDH is typically used as the internal control. A standard curve is made from a dilution series of a BCR/ABL-positive cell line, and the amount of residual leukemia cells is calculated by using this standard curve.

Advantages of real-time PCR over standard nested PCR for BCR/ABL include a decreased turnaround time, decreased chance for post-PCR contamination, decreased variability of results because the data collection occurs in the exponential phase of the PCR reaction, high throughput, and the possibility of obtaining quantitative results. Real-time PCR procedures are much more amenable to interlaboratory standardisation than nested PCR analyses. The major disadvantage of real-time PCR testing is the inability to compare the size of any detected rearrangements to that of the original malignant clone without additional testing.

However, this is not a problem with MRD testing in CML, as there is not a background population of normal cells carrying the BCR/ABL translocation. Quantitative real-time PCR technology can be used with many translocation targets, and can also be used for antigen receptor gene rearrangement analysis. The determination of the trend in the quantitative numbers of residual BCR/ABL-positive cells over a period of time is thought to provide important therapeutic information in the follow-up of CML patients. Optimal methods for quantitative real-time PCR detection of MRD in CML patients have not yet been established, and this testing is currently performed mainly in a few reference laboratories.

Once again, the importance of remembering the limitations of very targeted molecular testing in CML must be stressed. The clinician must be alert for development of clonal abnormalities in BCR/ABL negative cells or the development of resistance to imatinib mesylate.

The presence of mutations or amplification of BCR/ABL is known to be associated with resistance to imatinib mesylate in CML patients, and testing for these abnormalities is likely to become a standard part of the evaluation of CML patients in the future, at least for those who fail to achieve and maintain a CCR.

Follicular Lymphomas

The recent use of therapeutic modalities such as autologous bone marrow transplant following ex vivo purging of bone marrow B cells, monoclonal antibody therapy, and vaccine therapy has resulted in improved clinical outcomes of patients with follicular lymphomas (FL). PCR analysis for MRD performed on serial bone marrow samples in treated FL patients in complete remission has shown that some patients do achieve a molecular remission, and that the failure to achieve or maintain a molecular remission is predictive of relapse., Although the optimal methodology and timing for detection of MRD has yet to be determined, the t(14;18)(q32;q21)–IgH/BCL2 translocation, seen in 80%–90% of FL, is a good target for MRD detection. Unlike MRD testing in CML, in patients with FL, bone marrow analysis is clearly more sensitive for detection of MRD than peripheral blood. Many FL patients clear FL cells from the blood, while they still have persistent marrow involvement.

Nested PCR assays have been used historically and remain the most sensitive methodology for FL MRD detection; nested PCR can detect one translocation carrying cell in 10^6 normal cells and is still used in many laboratories for detection of this translocation. However, other molecular laboratories have switched from nested PCR to quantitative real time PCR. Real-time PCR for IgH/BCL2 is less labor-intensive and lacks the risk of contamination of standard nested PCR, but does have an analytical sensitivity that is usually at least 1 log less than that of nested PCR for IgH/BCL2. An additional and under-recognised problem in interpretation of real-time PCR analyses for MRD in FL is the potential for false positive results from occasional benign IgH/BCL2 translocation-carrying cells.

These are present in 30–40% of normal individuals, and with a highly sensitive test, a false positive result could occur if a comparison to the original FL clone is not made., This comparison is readily performed with nested PCR but would require substantial additional molecular analysis with real-time PCR. Using a quantitative real-time PCR method on serial bone marrows for MRD detection in FL, determining a trend over time, may obviate the necessity for this additional testing. Additional studies are needed to evaluate the clinical efficacy of MRD detection in FL patients in general, and to compare the relative clinical value of the nested and quantitative real-time PCR MRD detection methods.

An example of nested polymerase chain reaction (PCR) analysis for detection of minimal residual disease (MRD) in marrows from 2 follicular lymphoma (FL) patients.

Each sample is subjected to PCR for â-globin housekeeping gene and IgH/BCL2; PCR products undergo gel electrophoresis and (+) bands are detected by ethidium bromide staining (centre gels = â-globin and bottom gels = IgH/BCL2 MBR). A band is seen only if an appropriate product is detected; specificity of bands is confirmed by Southern blot using BCL2 probe (top autoradiographs).

(Panel A): Multiple controls are run in parallel with patient samples. (+) controls are serial dilutions of an IgH/BCL2 (+) cell line (lanes 4–7). The 10^6 dilution is run twice; as shown, a (+) result is often detected in only 1of these samples at the lower limit of analytical test sensitivity.

(Panels B andC): Parallel PCR analyses of 5 sequential marrows from 2 different FL patients are shown. For each patient, sample A was at diagnosis, before any treatment, B was 1–2 months after completion of CHOP chemotherapy but before treatment with anti-CD20 monoclonal antibody, and marrows C, D, and E were obtained at 2 months, 6 months, and 12 months after CD20 therapy. PCR of all clinical samples is run in duplicate. The patient in Panel B has a (+) nested PCR for IgH/BCL2 MBR at diagnosis and after CHOP, but the marrow becomes PCR (–) by 2 months after CD20 therapy and remains (–) at 6 and 12 months. The patient in Panel C is also PCR (+) at diagnosis and after CHOP, but also has a (+) band in 1 lane at 2 months after CD20 therapy. The (+) band is confirmed to represent a benign IgH/BCL2-carrying cell, not MRD, as it is not present in the duplicate sample and is clearly a different size from the patient's FL clone. Note that results are more easily interpreted on the BCL2-probed Southern blots than on the ethidium bromide–stained gels.

Precursor B-cell Lymphoblastic Leukemias

Multiple large prospective studies have clearly demonstrated the high prognostic value of MRD monitoring in children with precursor B-cell lymphoblastic leukemias (pre-B-ALL)., In childhood pre-B-ALL, studies of MRD have generally targeted patient-specific IgH antigen receptor gene rearrangements. This method takes advantage of the fingerprint-like sequences of the junctional regions of rearranged IgH genes, which differ in length and composition for each lymphocyte clone. To obtain these sequences, standard IgH PCR analysis is performed at diagnosis and/or relapse and the PCR products are Southern blotted, followed by sequencing of junctional regions of the clonal IgH rearrangements.

The different IgH rearrangements are then used for design of patient-specific oligonucleotide primers that are subsequently used in real-time PCR assays to follow the patient. Patient-specific IgH primers increase PCR sensitivity up to 1000-fold compared to standard consensus primers for IgV_H gene rearrangements; reactive background rearranged B cells do not obscure the clonal PCR products. At the present time, patient/clone-specific IgH PCR is not practical outside of a funded clinical trial setting, but this technique clearly offers the best potential for a sensitive, specific, and rapid analysis method that could be used over the course of therapy in most patients with pre-B-ALL.

Indeed, this same type of patient/clone-specific IgH PCR technology could be used for MRD detection in most other B-cell lymphomas (BCL) also, in which either no translocation-associated molecular event is available for MRD testing or the recurrent translocations occur in too low a proportion of the BCL subtype to be clinically useful. Standard IgH gene PCR with consensus rather than patient-specific IgH primers can usually detect only 1 malignant cell in 10^{2-3} normal cells, so the only sufficiently sensitive and specific method of testing for MRD detection in most BCL is therefore the use of patient/clone-specific Ig gene rearrangements. Quantitative real-time PCR techniques using standard IgH primers may provide some early information about the trend of the disease course over time, but will become negative when the patient could still have substantial residual disease. The combination of patient-specific IgH primers and quantitative real-time PCR could make a major contribution to the achievement of standardised MRD detection in BCL.

Conclusion

Whether offering or ordering a molecular test, the physician should know the circumstances in which the test should be ordered, the circumstances in which the test would not be useful, the advantages and the limitations of the test, and how to interpret the results. Many clinicians and pathologists are unfamiliar with molecular tests for hematologic malignancies, and misinterpret results of molecular testing. To avoid this, clinicians must be knowledgeable about the molecular test they are ordering and cautious about overinterpretation of results.

Physicians ordering molecular tests must be prepared to offer counseling on them, either personally or by referral. Good patient consent forms for molecular testing are crucial, and should explain to patients the meaning of a positive test, a negative test, and an inconclusive test. The consent form should inform the patient that the test could uncover other clinically relevant information, things that were not even being looked for.

Clinical molecular laboratories today are faced with two daunting tasks. First and foremost is the necessity of expanding test menus to meet the increasing clinical demand for testing for new genetic markers in hematologic disorders. Expanding test menus to meet clinical needs will require technical advances, as the current technology in clinical molecular laboratories does not allow rapid screening for a broad panel of relevant genes at a reasonable cost. However, equipment development has not been aimed at clinical laboratories, which traditionally have low budgets for new equipment, but at large pharmaceutical companies with abundant cash for new purchases.

This disconnect must clearly be addressed if clinical molecular testing is to be advanced, but even if the technological bottleneck preventing translation of new genetic knowledge to the clinical arena is alleviated, the lack of a reasonable level of reimbursement for molecular testing in general is still a major roadblock to successful implementation of new techniques for clinical molecular diagnostic testing. In most cases, the amount reimbursed for molecular diagnostic testing in hematopoietic malignancies in the United States today is inadequate to even cover the costs of test reagents. Little progress in the ideal model of "personalised" medicine will occur if this lack of funding persists.

Molecular Signatures of Lymphoid Malignancies: Identification of Novel Disease Subtypes and Rational Therapeutic Targets

Lymphoid malignancies are currently classified on the basis of morphology, immunophenotype, genetic features, clinical characteristics, and possible normal cells of origin. With the sequencing of the human genome and associated development of representative

DNA microarrays, it is now possible to obtain broad-based transcriptional profiles of specific lymphoid malignancies and previously unidentified disease subtypes.

The most commonly used platforms for gene expression profiling are cDNA and oligonucleotide microarrays With cDNA arrays, polymerase chain reaction (PCR) products of cDNA clones are spotted on filters or glass slides. A potential advantage of cDNA arrays is that they can be designed to address specific biologic questions. For example, the recently described "lymphochip" cDNA arrays are enriched for genes with documented importance in lymphocyte biology. Oligonucleotide microarrays include oligonucleotide probes deposited or synthesised directly on the surface of a silicon wafer. Oligonucleotide microarrays can potentially offer additional specificity by tailoring probes to reduce cross-hybridisation and discern splice variants. A common oligonucleotide array platform also facilitates comparisons across datasets of different tumour types .

Analysis of Gene Expression Datasets

One of the lymphoid malignancies in which gene expression profiling has been informative is diffuse large B-cell lymphoma (DLBCL). The most common lymphoid malignancy in adults, DLBCL comprises almost 40% of all lymphoid tumours. Although a subset of DLBCL patients can be cured with standard adriamycin-containing combination chemotherapy, the majority die of their disease. Robust clinical prognostic factor models such as the International Prognostic Index can be used to identify patients who are less likely to be cured with standard therapy. However, such models do not provide specific insights regarding more effective treatment strategies. For these reasons, additional insights into molecular bases for the observed clinical heterogeneity in DLBCL are critically needed. In addition, the multiple genetic abnormalities associated with subsets of DLBCL reflect additional molecular heterogeneity in this disease.

Investigators have utilised gene expression profiling to elucidate molecular bases for observed differences in DLBCL, identifying possible normal cells of origin, tumours with different responses to standard combination chemotherapy, novel rational treatment targets, and related disease entities (Savage et al, unpublished material). For these reasons, the lessons from gene expression profiling in DLBCL are likely to be broadly applicable to other lymphoid malignancies.

In one of the earliest applications of gene expression profiling, cDNA microarrays (lymphochips) and unsupervised learning techniques (hierarchical clustering) were used to characterise the transcriptional profiles of DLBCL and normal lymphocytes, including germinal centre (GC) B cells and in vitro activated peripheral blood B cells. In a pilot study, subsets of DLBCLs were found to share gene expression patterns with normal GC B cells or in vitro activated PB B cells. In an expanded analysis, a refined cell-of-origin signature (100 genes that distinguished GC-B-cell-like and activated-B-cell-like lymphomas at a significance level of $P < .001$) was used to identify tumours with features of above-mentioned normal B cells and a third unrelated subset.

Additional investigators have utilised supervised learning methods to develop transcriptional profiles of cured versus fatal/refractory DLBCLs. Genes implicated in outcome signatures included ones that regulate B-cell receptor signalling, critical serine/threonine phosphorylation pathways, and apoptosis. Two of the genes and pathways identified in this supervised outcome analysis have already been credentialed as possible rational therapeutic targets in DLBCL. In additional analyses, a combination of unsupervised and supervised learning methods were used to develop a DLBCL outcome model that included the cell-of-origin distinction and additional parameters including HLA class II expression and indices of proliferation.

These extremely powerful computational strategies provide new mechanisms for identifying discrete subsets of DLBCL and other lymphoid malignancies. The next challenges will be to link the molecular signatures of cell-of-origin and prognosis in lymphoid tumours with implicated biological pathways, specific pathogenetic mechanisms, and associated rational targets of therapy.

Proteomic Analysis of Hematolymphoid Neoplasms: Diagnostic, Biologic, and Therapeutic Implications

Hematolymphoid neoplasms are responsible for over 60,000 deaths annually in the United States, and are the most commonly occurring cancers in children. Despite these sobering statistics, it is within this field that molecular medicine has made its earliest and greatest strides, the promise of which is just beginning to be realised. The past few decades have seen the discovery of the t(9;22) BCR/ABL translocation in chronic myelogenous leukemia (CML), the characterisation of the role in apoptosis of the BCL-2 family of proteins, the use of microarray analysis to delineate new subsets of diffuse large B-cell lymphoma (DLBCL), and the introduction of novel biologic agents such as rituximab and imatinib, which already have had far-ranging impact in reducing the burden of cancer in selected patients. The field of hematolymphoid neoplasms remains fertile ground for the application of technology in the molecular diagnosis, characterisation, and treatment of human disease.

Overview of Proteomics

The functional effectors of cellular pathways and processes are proteins. While these proteins are encoded by the genome, only a subset of the possible protein products of the genetic code are produced, and the functional status of these proteins often depends heavily on post-translational modifications that are not reflected in their genomic sequences. Thus, while significant advances have been made from the analysis of the genome and its transcribed complement of mRNA, the protein end products of these processes are the effector arm of cellular events and offer an in vivo, functionally relevant window into the workings of the cell. The study of this wide complement of proteins derived from the genome is known as proteomics, and the proteins collectively are called the proteome.

The analysis of proteins is used daily in the clinical diagnosis and treatment of hematolymphoid neoplasms. An example is BCL-2 protein, an antiapoptotic protein overexpressed as a result of the t(14;18) translocation in follicular lymphoma. Detection of BCL-2 protein by immunohistochemistry is used routinely in the diagnostic evaluation of B-cell lymphomas. The presence of BCL-2 protein can yield prognostic information as well, such as in DLBCL, where BCL-2 protein expression has been correlated with decreased survival. Biologically, the elucidation of the mechanism by which the BCL-2 protein acts has led to the discovery of a large family of proteins related to apoptosis, with relevance to a wide variety of human diseases in addition to follicular lymphoma.

Finally, understanding the role of BCL-2 has led to molecular therapies such as antisense oligonucleotide approaches to lessen the antiapoptotic effects of this protein. Thus, clinicians may use a single protein to assist in diagnostics, prognostics, biologic understanding, and development of targeted therapies.

However, analysing proteins one-by-one is laborious, time consuming, and likely to miss critical biologic events due to the sheer number of existing proteins that could be assayed. New proteomic technology allows us to gain an overview of thousands of proteins simultaneously as a proteomic pattern, analyse the individual protein signalling pathways being utilised by neoplastic cells, characterise the neoplasm biologically, and select specific targeted treatment modalities, known as "personalised" molecular medicine.

Diagnostics

The power of molecular diagnostics has been demonstrated for hematolymphoid neoplasms perhaps more than in any other field. The ability to assay for the presence of characteristic translocations and detect clonal immunoglobulin gene rearrangements not only is used for diagnosis on a routine basis, but has helped support the classification of distinct disease entities, such as anaplastic large cell lymphoma. A major advance has been the use of large-scale gene expression analysis to develop genomic patterns or "fingerprints" to aid in the classification of hematolymphoid neoplasms. Alizadeh et al demonstrated the ability of microarray data obtained from diffuse large B-cell lymphomas to delineate distinct patterns of gene expression, which not only had biologic correlates in terms of the phenotype of the neoplastic cells, but also conveyed prognostic information beyond that obtained using the International Prognostic Index (IPI) alone.

Proteomic technology has now advanced to the point where proteomic patterns can be analysed as "fingerprints" in much the same way DNA array data have been analysed. Importantly, while conventional protein assays query individual protein biomarkers, proteomic pattern analysis uses complex bioinformatic tools to interpret spectra representing thousands of proteins to distinguish between clinical entities, such as the presence or absence of cancer. Unlike DNA microarrays, where target sequences must be available to be printed on the array, proteomic pattern analysis does not mandate identification or isolation of each protein comprising the overall pattern. This is important since the proteome is significantly more complex than the genome, incorporating the results of multiple alternative splicing variants and posttranslational modifications. While the genome is now thought to encode approximately 30,000 genes, the number of protein species may exceed 300,000.

Proteomic pattern analysis has been greatly facilitated by advances in high-throughput mass spectrometry, especially surface-enhanced laser desorption ionization time-of-flight (SELDI-TOF) analysis In SELDI-TOF, proteins from a patient sample (e.g., serum, urine, tissue lysate) are bound to a chip. After washing off unbound proteins and impurities, a matrix is applied that is subjected to photoactivation by laser energy. As proteins are desorbed by the laser they are launched as charged ions, and analysis of their time-of-flight (TOF) allows calculation of their mass-to-charge ratios (m/z). The spectrum of m/z ratios thus obtained is processed using self-learning bioinformatic pattern recognition software. Analysis of large cohorts of data using learning algorithms based on artificial intelligence approaches can allow the discrimination between two groups of samples ("supervised" learning, such as distinguishing the presence or absence of cancer), or to identify data clusters within a population set that may represent novel disease entities ("unsupervised" learning).

Overview of Mass Spectrometry

Patient proteins bound to a chip are irradiated by a laser, launching desorbed proteins as charged ions. Detection of the time-of-flight (TOF) of these ions yields a spectrum of mass-to-charge ratios (m/z). This spectrum, reflecting the protein constituents of the original sample, is then analysed by self-learning pattern recognition software and compared to previously encountered spectra.

An example of the power of this approach is the recent report of the ability to characterise the proteomic spectra derived from sera of women with and without ovarian cancer. SELDI-TOF mass spectra were used as a training set to develop an artificial intelligence algorithm that, when applied to blinded samples, could accurately identify 100% of patients with ovarian cancer and 95% of controls. This approach was sensitive enough to accurately identify all cancer patients, even those with stage I disease; specificity was demonstrated by including women with benign ovarian conditions in the control group, as well as finding the algorithm incapable of detecting the presence of cancer in sera from patients with prostate cancer. A recent improvement in mass spectrometry allowed correct identification of sera from ovarian cancer patients and controls with 100% sensitivity and 100% specificity.

The classification of hematolymphoid neoplasms has evolved from a system based on morphology alone to one based on cell of origin, the determination of which is aided by detection of lineage-specific protein markers using flow cytometry or immunohistochemistry. This classification also utilises detection of additional protein markers which are not lineage-specific but rather relate to the pathogenetic mechanism of the disease (e.g., BCL-2) or other clinically relevant genetic events (e.g., p53 mutations). In many cases, panels of antibodies are chosen in part due to limitations of the techniques employed, such as assaying cell surface molecules by flow cytometry, or ability of antibodies to detect antigens in paraffin-embedded tissue sections. The use of proteomic patterns representing tens of thousands of protein species may be an enormously powerful tool to complement the ongoing efforts to classify hematolymphoid neoplasms in ways that are biologically accurate and clinically relevant.

Additional ways that proteomic profiles might be used diagnostically include screening for minimal residual disease after treatment, screening for the development of neoplasia in high-risk populations (e.g., posttransplant), screening for transformation of a low-grade neoplasm into a high-grade one, and characterising/predicting responses to therapeutic interventions. For example, lymphoma cells in vitro have been shown to demonstrate characteristic proteomic patterns after exposure to chemotherapeutic agents, and characterisation of in vivo patterns may serve as an early predictor of response and/or toxicity.

Molecular Characterisation and Treatment

As previously mentioned, proteomic profiles can have important diagnostic and prognostic implications without reference to the individual proteins which constitute these profiles. Even at the single protein level, the protein CD20 has been used widely as a marker of B cells and a target for anti-B-cell monoclonal antibody therapy, although these uses are not primarily based on its specific cellular function. Clearly, however, the widespread analysis of the proteome will yield extensive data regarding function and utilisation of critical protein pathways, as was discussed for the BCL-2 protein, and is expected to have far-ranging implications for the identification of molecular targets for pathway-specific biologic therapy.

The ability to analyse the nature of the proteome in human tissues has been greatly facilitated by rapid developments in the field of protein microarrays. The power of this technique has been enhanced by the development of laser capture microdissection, in which subpopulations of cells, such as lymphoid follicles, can be isolated from tissue sections using a laser pulse. Protein lysates prepared from these samples then can be robotically applied in miniature dilution curves to a solid phase array with multiple other samples. Hundreds of replicate arrays can be generated and probed for the expression of a large complement of proteins using specific antibodies, including those that differentially recognise cleaved and/or phosphorylated forms of key signal transduction molecules. In this way, protein microarrays can identify the particular signalling pathways utilised by a population of neoplastic cells to tailor specific targeted therapy to modulate the function of these pathways.

One particularly attractive use of this technology is characterising the status of apoptotic pathways in hematolymphoid neoplasms. Apoptosis is an essential element for the normal development and maintenance of the immune system.

The development of a healthy immune repertoire is a highly stringent process, necessitating the elimination of most developing lymphocytes; conversely, the ability to avoid apoptosis is critical to the rapid expansion of quiescent lymphocytes in response to foreign antigen. Families of proteins related to apoptosis, such as the BCL-2 family, therefore include both proapoptotic and antiapoptotic members to aid in the homeostatic regulation of these immune processes.

Perturbation of these homeostatic mechanisms is exemplified by the t(14;18) translocation in follicular lymphoma, leading to overexpression of BCL-2. Further studies have indicated that antiapoptotic signals from BCL-2 are important in nonfollicular neoplasms as well, in which complex interactions among multiple BCL-2 family members appear to be critical. A downstream level of regulation is the inhibitor of apoptosis protein (IAP) family, which exerts its effect by inhibiting effector caspases.

Because multiple proteins and pathways modulate apoptosis, protein microarrays are ideally suited to assay these pathways and determine which are functionally active in a neoplastic cell population in order to select molecular targets for biologic therapy. A similar approach could be used assaying cell cycle regulatory genes, tumour suppressor genes, transcription factors, etc.

Application of protein arrays to apoptotic signalling pathways. Jurkat cells were treated with Fas ligand to induce apoptosis. Arrays were printed using cell lysates in serial dilutions and probed using an antibody specific to the cleaved (active) form of caspase 3. Increased cleaved caspase 3 is noted in the Fas ligand–treated cells (upper left). Hundreds of replicate arrays can be printed and probed with other antibodies; arrays then can be scanned, quantified, and compared.

Conclusion

Revolutionary technological advances in the field of proteomics have generated powerful new tools for the analysis of human neoplasia and cell signalling pathways. High-throughput mass spectrometry has generated proteomic patterns from patient samples that have had powerful implications in the diagnosis of several human malignancies, and this technology is ready to be applied to hematolymphoid neoplasms to aid diagnosis, monitor response, and refine disease classification. Coupled with a new generation of protein microarrays capable of characterising the functional status of multiple signalling pathways in neoplastic cells, these advances are leading to "personalised" molecular medicine, in which proteomic analysis will be used to select therapeutic agents to specifically target critical pathways in each individual neoplasm to optimise therapeutic efficacy and minimise toxicity.

Trials to validate the clinical use of these technologies are currently under way.

Chapter 5

Human Immunodeficiency Virus Hematology

The advent of potent antiretroviral therapy has altered the expected natural history of human immunodeficiency virus (HIV) infection and of many previously associated opportunistic complications, including malignancies. At the same time, HIV suppression hasn't affected all of these complications equally and the longer expected survival of infected patients may allow the development of newer complications. Additionally, the use of potent antiretroviral combination therapy may itself lead to hematological toxicities. Together these changes affect the consultation role of the hematology-oncology specialist in comprehensive HIV care and demand ongoing education.

In Section I, Dr. Paul Volberding reviews the biology of antiretroviral drug development and the progression in discovering new agents as the viral life cycle is further elucidated. He briefly summarises the process of combining agents to achieve the degree of viral suppression required for long-term clinical benefit.

In Section II, Dr. Kelty Baker reviews the effects of HIV and its therapy on hematologic dyscrasia and clotting disorders. She summarises how therapy may decrease certain previously common manifestations of HIV disease while adding new problems likely to result in referral to the hematologist. In addition, she addresses the role of secondary infections, such as parvovirus, in this spectrum of disorders.

In Section III, Dr. Alexandra Levine discusses the still challenging aspects of HIV associated non-Hodgkin's lymphoma and the association between HIV infection and Hodgkin's disease. She addresses current controversies in the pathogenesis of HIV related lymphomas and summarises a number of recent trials of combination chemotherapy, with or without monoclonal antibodies, in their management.

Additionally, she reviews the complex relationship of HIV disease with multicentric Castleman's disease and recent attempts to manage this disorder.

The Pathophysiology of HIV Therapy

The HIV epidemic, the worst epidemic in human history, has already taught us much. It is a convincing illustration of the balance between the power of science and the humanism of modern medicine, a relationship that further integrates behavioural medicine with clinical care. The epidemic has also given birth to an invaluable and informed patient advocacy community—one with clear implications for other diseases.

Antiretroviral therapy, remarkable in its clinical benefits, is itself a paradigm of the dramatic pace of molecular medicine. In only a very few years, HIV therapy moved from limited to striking potency, an advance based on rapid discovery of the structures and functions of the HIV virion and life cycle. As the same tools that elucidate HIV biology are applied more broadly in medicine, it is possible to be optimistic that other diseases will be similarly deciphered and controlled. Drug discovery based on fundamental molecular pathogenesis is rapidly reshaping our concepts of medicine from those based on organ-specific diseases to those recognising common disease mechanisms. Within a given area, research in one disease illuminates others. For example, advances from HIV research are being applied to new challenges. Emerging infectious diseases such as West Nile Virus and now severe acute respiratory syndrome (SARS) have been quickly characterised and drug discovery is moving forward using approaches in many cases developed in response to HIV/AIDS.

This review will address the biology of HIV as it relates to antiretroviral treatment. It will summarise current questions and approaches in HIV management focusing primarily on the virology of HIV infection. Intentionally, it will not address the immunology of HIV infection in any detail, not because it is less important, but because, to date, it has fewer therapeutic agents of established efficacy. The review will also pay only passing attention to the behavioural components of HIV therapy, such as medication adherence. These are critically important—uniquely so in HIV medicine because of rapid selection of drug resistance mutations—but are beyond the scope of this review. This review will refer liberally to several excellent reviews of HIV virology and attempt to summarise the clinical application of this new knowledge.

The Origins of the HIV Epidemic

The virus responsible for essentially all AIDS cases in the Western world, HIV-1, represents human infection by a virus originating in another species, in this case, the chimpanzee (which may, in turn, have been infected by another primate species). Based on mathematical modelling of HIV sequence variations, it is estimated that the initial human infections occurred near 1930 in sub-Saharan Africa. The HIV epidemic was probably accelerated by urban population concentration in postcolonial Africa.

It was spread worldwide by rapid transportation. Such zoonotic transmission events are now appreciated as quite common. Recent examples include outbreaks of HIV-2, SARS, and monkeypox.

HIV-1, hereafter more simply termed HIV, collectively includes a series of similar but unique and relatively stable sequence variations called clades. Each clade is thought to represent an original animal to human transmission event. The biologic variation between clades is an area of very active investigation and may prove clinically relevant, affecting pathways of resistance selection. HIV clades are useful as "fingerprints" in tracking the worldwide HIV spread and local prevalence of one clade is quite characteristic and distinctly different than another region.

For example, while clade B is almost universal in infected persons in the United States, clade C is similarly dominant in southern Africa. Regions at the interface of two dominant clade prevalences may also contain a substantial population infected with recombinants of the two clades. It is not clear to what degree this represents simultaneous dual infection or later superinfection of a new clade in a patient already chronically infected with a first clade. True superinfection, if common, could suggest difficulties in vaccine development as it would indicate the failure of host response even when ideally primed.

HIV Infection as a Disease

Initial cases of advanced immune deficiency (almost immediately associated with CD4 T-lymphocyte depletion), often with striking opportunistic infections or malignancies, led to the recognition of the AIDS epidemic. The identification of HIV as the causative agent offered insight into the epidemiology and pathogenesis of AIDS. Tests for HIV-induced antibodies found a large epidemic, mostly of those not yet ill but with signs of progressing immune depletion. Clinical attention thus appropriately shifted from AIDS to HIV and the recognition that HIV infection is a disease in its own right and one, moreover, appropriate for treatment.

HIV infects CD4 cells and can be directly cytotoxic. Other complex mechanisms, however, appear important in explaining the progressive depletion of this key cellular element of immunity. Immune-mediated cytotoxicity is probably involved, as T-cell regeneration in the thymus is also limited by HIV infection. Along with CD4 T cells, macrophage/ monocytes are also infected. These cells may be key to maintaining and spreading HIV within the body and are probably not directly killed by infection. Along with $CD4^+$ cell depletion, which is the easiest measure to stage HIV disease, infection damages many components of normal immunity. T-cell proliferative response, especially to HIV antigens, is quickly lost and the repertoire of antigen recognition diminishes with time.

HIV Structure

HIV is a small virus containing 2 single RNA strands in a core structure bound by a gag-derived protein, p24. The viral core also packages key enzymes needed for replication, including reverse transcriptase, integrase, and an HIV specific protease. These will be described in more detail below.

The HIV core is contained within a lipid bilayer, a portion of the infected cell membrane taken during viral budding. The viral envelope contains specific structures, several key to infectivity. One, a glycoprotein of 120,000 molecular weight, gp120, was quickly recognised as the main viral structure interacting with the CD4 protein on the surface of cells susceptible to infection. Most early attempts to create an HIV vaccine used recombinant gp120 as an antigen in the hope that antibodies to this protein would effectively neutralise HIV, a strategy unsuccessful to date. Although neutralising antibodies are found in some cases, their low titer and narrow specificity made these antibodies incapable of providing protection against the variable viruses in actual infection.

The glycoprotein gp120 is bound to the HIV envelope by a second glycoprotein, gp41. gp41 has an envelope spanning region, but also a complex triple-helical region projecting from the outer aspect of the envelope. This, along with gp120, appears to be essential for cellular infectivity.

As infection necessarily involves interactions between the virus and the target cell, cellular membrane structures are also involved.

These include the CD4 antigen and, as will be described below, chemokine receptors, particularly CCR5 and CXCR4.

The HIV Life Cycle and Antiretroviral Drug Development

Many elements of the HIV life cycle have been elucidated, at least sufficiently to provide a basis for antiviral drug development. Major steps in this life cycle will be reviewed, followed by a brief discussion of antivirals available or in development targeting that biology.

Binding of gp120 to CD4

One of the domains of CD4 binds to a structurally protected pocket of the gp120 viral glycoprotein. Even simple mutations in this gp120 site affect binding significantly. The CD4/gp120 interaction is necessary, but not itself sufficient for cellular infection. As a key step in infection, however, numerous attempts have targeted this for antiretroviral therapy. In early and unsuccessful approaches, the small, water soluble, gp120 binding domain of CD4 was cloned, expressed, and injected in HIV-infected subjects. This soluble CD4 was designed to occupy the gp120 of HIV and was active in vitro. It was inactive, however, against the more variable HIV of actual primary isolates. Recently, one company announced early development of a small molecule inhibitor of gp120/CD4 binding, a promising approach that could potentially avoid the toxicities seen with drugs acting within the cell.

CD4 Undergoes Structural Transformation Following gp120 Binding

This "melting" of CD4 allows a subsequent interaction between gp120 and the chemokine receptors physically close to CD4 at the cell surface. One company has developed a humanised murine anti-CD4 monoclonal antibody that binds to the pretransformed CD4, stabilising it. This prevents HIV infection of the cell, presumably by preventing gp120/chemokine receptor interaction. This antibody was effective in vitro and in an early Phase I trial in human HIV-infected subjects. Individuals were given a single intravenous dose of the monoclonal antibody. Higher doses resulted in a 1–1.5 log viral suppression lasting for as long as 3 weeks with no reported toxicity or antibody production. Multidose trials are currently in process.

The gp120 Interacts with the Chemokine Receptor

This step in the viral life cycle is another target of active drug development. Almost all patients are initially infected with a virus that uses the CCR5 chemokine receptor, often termed R5 HIV. Some cases of advanced HIV disease show evolution of the virus (through mutations affecting the gp120 binding site) to enable use of the CXCR4 chemokine receptor, termed X4 HIV. There is some evidence that the X4 virus is more rapidly cytopathic than R5 and thus concern that blocking interactions with the CCR5 chemokine receptor may favour evolution to this more virulent form.

To date, however, this has not been observed in drug development. Various strategies of blocking the gp120/chemokine receptor are being explored, and this approach does appear amenable to small molecule drug development. In at least 1 case, a promising compound was found to prolong the Q-Tc interval, a potentially serious toxicity.

Other related compounds that do not share this problem are in development. One question in chemokine blocking is whether this will cause impairment of one component of innate immunity, the natural function of these structures. To date, this has not been observed. Perhaps speaking to this issue is the observation that individuals with a homozygous deletion of the gene coding for the CCR5 chemokine receptor show no evidence of immune deficiency, although they are naturally resistant to infection with R5 HIV.

The gp120–chemokine Receptor Interaction Triggers Uncoiling of the Triple-helical External Domain of HIV gp41, which is then Inserted into the Cell Membrane Leading to Fusion of the Viral and Cell Membranes

This step, an elegant biologic process, has been resolved through careful structural biologic investigation. Understanding this process has already resulted in successful drug development with the approval of enfuvirtide. In a manner comparable to that of a harpoon, following the uncoiling of gp41, the triple-helical structure extends and inserts its tip containing the fusion domain directly into the cell membrane. This tethers the virus to the cell.

Fusion and actual infection require a subsequent recoiling of the gp41, an active process that approximates the 2 membranes. This recoiling is prevented by enfuvirtide (formerly T-20), a 36-mer polypeptide that binds to the uncoiled, triple-helical region of gp41, stabilising it in that conformation.

Because of its novel mechanism, enfuvirtide is fully active, despite high-level resistance to all other antiretroviral drugs. The drug must be given as a twice daily subcutaneous injection. If not combined with other active antiretrovirals, resistance develops quickly. Other fusion inhibitors are also being developed. One, T-1249, acts in HIV that has developed enfuvirtide resistance and can be given as a single daily subcutaneous injection.

Reverse Transcription of the Viral Single-strand RNA Eventuating in a Dual Strand DNA Copy Occurs in the Cytoplasm Following Viral Entry

Reverse transcriptase, the enzyme needed for this process, is contained within the viral core and coded by the HIV genome, *pol*. Reverse transcription, relatively unique to the lentiviruses, including HIV, was an obvious and quickly successful target of drug development. Already, 2 classes of reverse transcriptor inhibitors are in use: 8 nucleoside-analog inhibitors and 3 with other structures (thus termed non-nucleoside reverse transcriptase inhibitors). In both categories, new agents are in advanced development stage.

In common parlance, the reverse transcriptase inhibitors structurally related to nucleosides or, in one case, nucleotides, are collectively called the nucleoside reverse transcriptase inhibitors (nRTIs).

The nRTIs are often considered the “backbone” of HIV therapy and, very typically, 2 of these drugs are included in multidrug regimens.

Appreciating that full medication adherence is essential to durable clinical benefits, several coformulations of these drugs are either available or in development, thus reducing daily pill burden. The nRTIs have few class-specific toxicities, but each agent has a unique and important toxicity profile. Side effects range from mild to life threatening.

Human Immunodeficiency Virus (HIV) Drugs

Drug resistance is key to regimen design. While certain drugs select quickly for a single mutation conferring high-level phenotypic resistance,, others require multiple mutations over prolonged time for similar consequences.

Furthermore, it is believed that certain resistance mutations may compromise the capacity of the virus to replicate compared to wild-type HIV and that such mutations may allow some continued clinical benefit, even in the face of high-level resistance. This observation is leading to very interesting treatment strategy trials in advanced HIV virologic failure. This is, again, beyond the scope of this review.

Reverse Transcriptase Inhibitors not Structurally Related to Nucleosides are Termed Non-nucleoside Reverse Transcriptase Inhibitors (nnRTIs)

The nnRTIs are more alike in terms of toxicity and resistances than their nRTI relatives. These agents bind to a very specific site on the reverse transcriptase. They are inactive against the otherwise closely related virus HIV-2, and a single genotypic mutation is able to bestow complete viral resistance. This resistance is essentially complete across the group of currently approved nnRTIs. Thus, in contrast to the nRTIs—where continued activity is possible with selective, sequential replacement of drugs within the class with the nnRTIs, only a single failed attempt is possible. For this reason, these drugs require careful selection of supporting agents, careful patient selection, and scrupulous adherence.

Despite seeming limitations, the nnRTIs are crucial components, especially in early treatment regimens. Two nnRTIs, nevirapine and efavirenz, are in common use. A third, delavirdine, is less convenient, less potent, and is rarely chosen. The nnRTIs, as a class, are potent and the most widely used are convenient with compact formulations and prolonged half-lives enabling once- or twice-daily administration. Toxicities of nnRTIs tend to be most common during initial weeks of use. Rash and hepatic effects are seen with both commonly used agents. Central nervous system side effects are common with efavirenz, while hepatotoxicity is more common with nevirapine, especially in patients coinfected with HIV and either hepatitis B or C. Newer agents are in development in this class, some appearing to have unique resistance pathways.

The DNA Copy of the HIV Genome is Integrated into the Host Cell DNA in the Nucleus Through the Action of the HIV Integrase Enzyme

This process is quite complex but substantial advances in our understanding of it are leading to early drug development. Among the steps in the integration process are the interactions in the cytoplasm (again, integrase is carried into the cell within the viral core) where the enzyme is attached to each end of the newly copied viral DNA, forming a complex that is then taken into the nucleus through pores in the nuclear membranes. The integrase then breaks the host DNA probably in selected sites and transfers the viral DNA strand to the opened end of host DNA. Integrase then repairs the breaks, leaving the proviral DNA inserted into the host genome. Because the active site of integrase involves DNA, the structure has been difficult to resolve. But it is now relatively well characterised and is the basis of active drug development. Several compounds have shown in vitro activity and at least one is in early human trials.

The Proviral DNA is Activated and Transcribed into RNA, Which is then Translated to form Viral-coded Proteins

Viral activation typically occurs early after infection, but in some cells does not occur even after prolonged periods. These latently infected cells are thought to form a reservoir of infection, making viral eradication or cure highly improbable, as most antiretroviral drugs target active steps of replication. The activation and transcription of HIV, and the transport of gene products from the nucleus to the cytoplasm, involves regulatory genes *tat* and *rev*. To date, drug development of agents targeting these genes or their gene products have been unsuccessful; but they represent attractive areas of investigation as interfering with them may prove useful in decreasing the production of viral particles from already infected cells.

HIV viral components assemble at the inner aspect of the cell membrane, bud from the cell, and finish maturation into an infectious virion after release.

A very active area of HIV research and drug development addresses the later steps in infection. While some aspects such as the selective proteotic cleavage of precursor HIV proteins is well understood and has led to important approved agents, earlier steps in viral assembly are only now being elucidated. It is clear that viral assembly is carefully targeted. Certain regions of the cell membrane are favoured for viral assembly and budding, and cells contain endogenous materials that inhibit retrovirus production. Still under active investigation, lentiviruses including HIV have a gene, *vif*, that appears to inactivate this "natural" protective cellular mechanism through an interaction with the product of the *APOBEC3G/CEM 15* gene. An intriguing potential target for drug development, this system is being explored by several HIV laboratories.

Once the assembly of the virus inside the cell membrane is advanced, the budding begins to pinch off the forming viral core within a pocket of the cell membrane. The immature viral particle is at first attached to the cell by a thin stalk of cell membrane, which then breaks to release the virion into the external environment. The process of budding and release is still rather poorly understood, but is a likely target of drug development.

As the virion buds and is released, HIV-specific protease begins to cleave viral proteins into active truncated forms needed for full viral maturity and infectivity. Specific inhibitors of HIV protease are in wide use .Their potency has revolutionised HIV therapy and the continued development of new compounds is addressing problems seen with the first generation of this drug class. Very broadly speaking, early protease inhibitors (PIs) were inconvenient, with high pill burden, large pill size, and restrictions on use with food and water. As a class, these agents were often associated with gastrointestinal disturbance and dyslipidemias.

Specific protease inhibitors had other toxicities limiting use in some patients. Resistance to this drug class resembles the nRTIs more than the nnRTIs, typically requiring multiple mutations for full resistance and usually occurring only after prolonged viremia in the presence of the drug. The cytochrome p450 induced metabolism of most members of the class is significantly blocked in the presence of even low concentrations of ritonavir, one of the earliest of these drugs to be developed. This effect has enabled more convenient dosing of this class.

Ritonavir, while poorly tolerated in full doses, and thus seldom used as a protease inhibitor in its own right, is often used in conjunction with a second protease inhibitor to boost that drug's pharmacologic profile. An interaction seen with all drugs in the class except nelfinavir, low-dose ritonavir raises peak and trough concentrations of protease inhibitors as well as their integrated concentration over time, the area under the curve (AUC). Ritonavir enhancement or boosting makes it possible to use indinavir twice daily instead of 3 times daily and enables other protease inhibitors to be used once daily. Ritonavir enhancement is necessary for the investigational protease inhibitor, tipranavir, to be used at all, given otherwise limited serum levels.

Protease inhibitor resistance is a complex problem and the subject of much research. Mutations conferring resistance may well impair viral fitness similar to that observed with the nRTIs. Protease resistance patterns may have implications for the choice of initial and salvage protease inhibitor–containing regimens. One recent observation suggests that certain resistance mutations may even confer increased susceptibility to other members of this drug class.

Protease toxicity is quite drug specific and beyond the scope of this review. But it is clear that some agents in this class may interfere with both carbohydrate and lipid metabolism. Highly controversial and still being clarified in prospective cohort studies, protease inhibitor use may contribute to visible and characteristic morphologic alterations in lipid deposition, although recently some nRTIs have been even more implicated, especially in the loss of subcutaneous fat. Hyperlipidemias including LDL cholesterol and triglycerides are seen with some protease inhibitors and have raised concerns in the field about accelerated cardiovascular complications. Carbohydrate metabolism effects cause relative insulin resistance and in some cases, frank diabetes.

Newer protease inhibitors are showing promise in many respects. New drugs or reformulations of existing ones allow similar and fewer pills and less frequent administration. One recently approved drug, atazanavir, can be used once daily, soon as a single daily tablet, and furthermore, does not cause hyperlipidemia.

General Approaches to HIV Therapy

Successful antiretroviral therapy allows durable suppression of viremia below the quantitation limits of sensitive viral assays. Treatment is best started before very advanced disease stages, but even when begun later can often substantially reverse the immune deficiency of HIV infection. The optimum timing of therapy and choice of drugs is well described in published treatment guidelines. In general, treatment should be initiated before a CD4 cell count of 200/mm^3, as the risk of complications, infections, and cancer increase below that level. Therapy can safely be deferred in most cases until the CD4 cell count is 350/mm^3, as the risk of symptomatic disease is small in that region and given the cost and adverse consequences—toxicity and resistance—of more prolonged therapy. An absolutely key issue is the patient's motivation to be treated and his or her ability to adhere fully to the prescribed regimen. Any delay in initiating HIV therapy should be used to prepare patients to comply once it has begun.

The choice of drugs in the initial regimens, given the almost limitless number of permutations of existing agents, is also complex, but recent guidelines from the Department of Health and Human Services (DHHS) have selected certain combinations as preferred.

Human Immunodeficiency Virus (HIV) Drug Regimens

But nearly as important is the individualisation of regimen choice in terms of convenience and side effects to improve the durability of clinical benefits and forestall nonadherence leading to drug resistance selection.

The field of HIV medicine continues to evolve rapidly. Vigorous fundamental research is constantly opening new possibilities for effective drug development and new insights into drug resistance and toxicity. Lessons learned from this investment in HIV biology are increasingly applicable to other fields of medicine and the tools of this science are immediately applicable to even newer pathogens introduced into the human population.

Anemia

Depending on the study setting, anemia can be found in 63%–95% of those with HIV infection at some point during the course of their disease. Sullivan and colleagues in the Multistate Adult and Adolescent Spectrum of HIV Disease Surveillance Project reviewed the records of 32,867 HIV-infected individuals and reported that the incidence of anemia increased with the clinical stages of disease. Three percent of those with HIV infection alone developed anemia at 1 year of follow-up, compared with 12% of those with immunologic AIDS (CD4 count less than 200/iL or CD4 percentage less than 14%) and 37% of those with clinical AIDS. The presence of anemia in an HIV-infected patient is significantly associated with an increased risk of death, and this is independent of the CD4 count or viral load. Conversely, recovery from anemia is associated with a decreased risk of death.

The most frequent cause of anemia in HIV-infected patients is anemia of chronic disease. $CD34^+$ stem cells express low levels of CD4 and CD11a, making them relatively resistant to direct infection by HIV. However, mononuclear-macrophage cells can develop productive HIV infection and the resultant release of cytokines, such as transforming growth factor-beta (TGF-â), tumour necrosis factor-alpha (TNF-a), and interleukin-1 (IL-1), contribute to the suppression of hematopoiesis. Although serum erythropoietin levels can be high in these patients, the elevation is blunted compared to patients with uncomplicated iron-deficiency anemia of comparable severity.

Anemia

Drug therapy for HIV infection or its subsequent complications is also a common cause of anemia. Zidovudine (AZT), the most notorious culprit, inhibits in vitro erythroid colony formation in a dose-dependent manner. Severe anemia, defined either as a hemoglobin level of less than 7.5–8.0 grams per deciliter or anemia that requires transfusion, can be seen in 24% of those who receive AZT 1500 mg daily. Although macrocytosis develops within weeks in most patients taking AZT and is a useful marker of compliance, neither vitamin B_{12} or folinic acid are helpful in preventing AZT-induced myelotoxicity. In fact, serum vitamin B12 levels are often low in HIV-infected patients, but only a few patients have true vitamin B_{12} deficiency, as indicated by elevated homocysteine and methylmalonic acid levels.

Other drugs that cause anemia in AIDS patients include ganciclovir, which can induce marrow suppression and pancytopenia. Amphotericin B may cause anemia by suppressing erythropoietin production, but this is controversial.

Trimethoprim interferes with folate metabolism in erythroid cells, while sulfamethoxazole, dapsone, and primaquine can cause hemolysis in glucose-6-phosphate dehydrogenase (G6PD)–deficient individuals. Survival in G6PD-deficient patients is, however, not compromised despite putative oxidative stress on red cells produced by HIV infection.

Human parvovirus B19, a small single-stranded DNA virus, can cause severe, chronic anemia in AIDS patients. The virus, usually acquired via the respiratory tract, invades erythroid progenitor cells via the blood group P antigen, and then replicates extensively, ultimately lysing the infected cell. Because of incompetent humoral immunity during advanced stages of HIV infection, patients are unable to generate neutralising immunoglobulin (Ig) M or IgG antibodies; consequently, viremia persists unchecked. The result is pure red cell aplasia, with a block in erythroid maturation and characteristic giant pronormoblasts found in the bone marrow.

These findings are not universal, however, so bone marrow aspiration should not be the cornerstone of diagnosis. Instead, the patient's serum should be assayed for parvovirus B19 DNA using dot blot hybridisation or polymerase chain reaction (PCR) assays. According to data from the University of Washington, evidence of parvovirus B19 infection is found by dot blot hybridisation in 17% of HIV-infected patients who have severe anemia.

Treatment consists of one or more courses of intravenous immunoglobulin therapy (400 mg/kg/day for 5 days) in an attempt to provide the antibodies necessary to clear the infection, but this is not efficacious in all individuals. Multiple recent case reports suggest that institution of highly active antiretroviral therapy (HAART), with subsequent reconstitution of the patient's humoral immunity, results in spontaneous clearance of parvovirus B19 infection.

Mycobacterium avium complex (MAC), *Mycobacterium tuberculosis* (MTB), and *Histoplasma capsulatum* cause anemia in AIDS patients by infiltrating the marrow. Patients from the Mediterranean may also develop anemia as a result of infection with *Leishmania donovani*, whereas those from Southeast Asia and southern China are susceptible to *Penicillium marneffei* infection. Bone marrow aspiration and biopsy provide evidence of infection in 25%–42% of patients tested.

Frequently, however, the diagnosis can be made equally rapidly and accurately using less-invasive modalities such as lysis-centrifugation blood culture, serology, or nucleic acid hybridisation. This is especially true for mycobacterial infections. Bone marrow provides a unique diagnosis in only 10% of the procedures performed, and is more likely to be useful in patients with low CD4 counts or a hematocrit less than 25%.

Surprisingly, autoimmune hemolytic anemia (AIHA) is not common in patients with HIV infection. Although the prevalence of a positive direct antiglobulin test (DAT) is high in this population, ranging from 18% to 43%, only a few cases of HIV-associated AIHA have been reported. DAT-positive patients generally have lower hemoglobin levels than DAT-negative patients. This is, however, most likely because both the prevalence of DAT-positivity and the severity of anemia increase with more advanced stages of HIV infection. In the few cases of symptomatic autoimmune hemolytic anemia reported in HIV-positive patients, both warm and cold antibodies have been found, sometimes simultaneously. Reticulocytopenia is sufficiently frequent in those with HIV infection that its presence cannot be used to exclude hemolysis.

However, the patient's haptoglobin level is usually decreased, the lactate dehydrogenase level elevated, and the bone marrow normocellular with erythroid hyperplasia. Successful treatments have included corticosteroids, intravenous immunoglobulin, withdrawal of any offending drugs, and splenectomy. Aggressive transfusion therapy should be undertaken with caution in HIV-associated AIHA, as fatal pulmonary embolisation due to augmented hemolysis and disseminated intravascular coagulation has been reported., Indinavir, ceftriaxone, and "Ecstasy" have been reported to cause hemolytic anemia in HIV-infected patients. Rarely, cytomegalovirus (CMV) infection can also cause hemolysis.

Fatigue, the cardinal symptom of anemia, is also the most common symptom of HIV infection and is responsible for impaired physical function and poor quality of life. Numerous open-label and randomised, double-blind, placebo-controlled trials using a variety of regimens have found that the use of recombinant human erythropoietin in HIV-associated anemia frequently results in higher hemoglobin levels, fewer transfusions, and a better quality of life. In an observational study at the Johns Hopkins HIV Clinic, treatment of anemic patients with erythropoietin also resulted in a significantly reduced risk of death.

In contrast, several studies have suggested that transfusion therapy, even early in the course of HIV infection, is associated with increased mortality., Explanations for this include the following: transfusion-associated infection with agents such as CMV, hepatitis B and C, human T-cell lymphotropic virus (HTLV I/II), or parvovirus B19; transient activation of HIV expression; and transfusion-associated immunosuppression mediated by inflammatory cytokines that cause decreased lymphocyte, natural killer cell, and monocyte function. However, a randomised, double-blind trial comparing the use of leukoreduced versus unmodified red blood cells in patients with advanced HIV infection found no evidence of HIV, CMV, or cytokine activation following transfusion.

Furthermore, leukoreduction provided no clinical benefit and may have even worsened survival. Excessive transfusion can cause iron overload, which may provoke HIV progression and increase susceptibility to infection with *Candida* species, *Pneumocystis carinii*, and *Mycobacterium* species by impairing macrophage function.

Thrombocytopenia

Dominguez and coworkers studied platelet kinetics in 41 HIV-infected thrombocytopenic patients and found that platelet survival was lower in those with CD4 counts above 200 cells/mL than in those with counts below this level, implying that platelet destruction is more important in patients with high CD4 counts, and decreased platelet production is more important in those with lower CD4 counts.

Similar kinetic studies performed by Cole et al concluded that HIV-infected patients have ineffective delivery of viable platelets to the peripheral circulation, despite a 6-fold elevation in thrombopoietin levels and a 3-fold expansion of megakaryocyte mass compared to normal controls. This finding suggests the possibility of HIV-induced apoptosis of megakaryocytes and is compatible with the results of kinetic experiments, which found increased platelet turnover but no change in platelet survival following the initiation of zidovudine (AZT) therapy, indicating that platelet production had increased during treatment.

Megakaryocyte infection by HIV is supported by the following: denuded nuclei and ballooning of the peripheral zone of megakaryocyte cytoplasm have been observed by electron microscopy; internalisation of HIV particles has been seen in coculture studies; the presence of the HIV p24 antigen has been shown by immunohistochemical techniques; and expression of HIV RNA has been found using in situ hybridisation. Marrow infiltration by infectious organisms or neoplasms, as well as adverse drug effects, can also cause impaired platelet production and thrombocytopenia.

Although 8% of patients with HIV-associated thrombocytopenia will have a hemorrhagic event, treatment is usually not necessary unless the platelet count is below 30,000/iL or the patient is symptomatic. Patients with hemophilia or other coagulopathies should probably receive therapy when their platelet counts are below 50,000/iL because of their higher risk of bleeding. As many as 18% of patients who have HIV-associated thrombocytopenia will undergo spontaneous remission. In those who do not, therapy historically has consisted of institution of AZT. Recent studies indicate that HAART is equally effective.

Other treatment modalities specifically for HIV-associated ITP include glucocorticoids, intravenous IgG (IVIG), intravenous anti-D therapy, splenectomy, danazol, interferon, and vincristine. Glucocorticoids, typically prednisone 1 mg/kg daily, increase the platelet count in many patients; however, long-term use can result in Cushing's syndrome, an increased risk of fungal infection, and acceleration of the course of Kaposi's sarcoma.

Infusion of IVIG induces rapid but unsustained remissions in 71%–100% of HIV-infected patients, but is costly and cumbersome to administer. Intravenous anti-D therapy is less expensive, but increases the platelet count above 50,000/iL in only 34% of patients treated. Although the response to anti-D is longer than that seen with IVIG, the extent of hemolysis in D^+ (Rh^+) patients is unpredictable. Patients with baseline hemoglobin levels above 12 g/dL are more likely to have a clinically important elevation of their platelet counts than those with anemia. Splenectomy can also be successful and, despite early concerns, does not obviously increase the risk of progression of HIV infection to symptomatic AIDS. In fact, a long-term cohort study of

45 patients (17 had splenectomies and 28 did not) demonstrated a significant reduction in the risk of developing full-blown AIDS and a trend toward reduced mortality in splenectomised patients. This may be due in part to a temporary reduction in plasma viremia and an increase in absolute CD4 and CD8 counts. Splenic irradiation is of minimal benefit, resulting only in small increases in platelet counts for brief periods of time.

Neutropenia

Causes of low absolute neutrophil counts in HIV patients include inhibition of granulopoiesis by the virus itself, marrow infiltration by infectious organisms or neoplasia, adverse drug effects, autoimmune neutropenia, and hypersplenism. Treatment of neutropenia includes initiation of antibiotic therapy for fever or other overt evidence of infection; antiretroviral therapy for individuals who have untreated HIV infection; withdrawal of potential offending drugs; and stimulation of myelopoiesis using either granulocyte colony stimulating factor (G-CSF) or granulocyte macrophage colony stimulating factor (GM-CSF).

Thrombosis

Thrombosis reportedly occurs in up to 2% of HIV-infected patients. Factors associated with venous thromboembolic complications include age over 45 years, advanced stage of HIV infection, the presence of CMV or other AIDS-defining opportunistic infections, hospitalisation, and therapy with indinavir or megestrol acetate. The association between opportunistic infections and thrombosis may simply reflect immobility due to illness. Alternatively, CMV may promote adhesion of neutrophils and platelets to the endothelium, induce production of antiphospholipid antibodies, enhance synthesis, and increase secretion and survival of factor VIII and von Willebrand factor. Why indinavir would predispose to thrombosis is unclear, but megestrol, like other progestational agents, may cause acquired resistance to activated protein C.

In addition, individuals with HIV infection may be at increased risk for thrombosis because of decreased levels of antithrombin, free protein S, protein C, or heparin cofactor II; the presence of anticardiolipin antibodies; coexistence of malignant, inflammatory, or autoimmune disorders; or vascular damage due to injection drug use, placement of intravenous catheters, or CMV infection. Antithrombin deficiency can occur in association with HIV nephropathy as a result of losses in the urine. The nephrotic syndrome seen in HIV nephropathy may also result in compensatory hepatic synthesis of factors V, VIII, and X induced by hypoalbuminemia, and increased platelet adhesion and aggregation. Acquired protein S deficiency can be found in up to 75% of HIV-infected children and adults, especially in patients with CD4 counts below 200/iL or AIDS, resulting in thrombotic complications in as many as 12%. Acquired free protein S deficiency in HIV is not caused by changes in levels or function of C4b-binding protein but rather by the appearance in some patients of antibodies against protein S. The levels of free protein S antigen in HIV patients can appear artificially low when assayed by the PEG precipitation technique, so that the prevalence of protein S deficiency in HIV-positive patients may actually be lower than previously thought (about 10%).

The lupus anticoagulant is found in 0%–70% of patients with HIV infection, depending on the sensitivity of the assay and the characteristics of the patients examined. Anticardiolipin antibodies are detected in 46%–90% of these patients. Neither is strongly associated with an increased risk of venous thromboembolic complications in HIV, but events such as multiple transient ischemic attacks and stroke, avascular necrosis of bone, skin necrosis, and brachial artery thrombosis have occasionally been described in patients with anticardiolipin antibodies.

Hemophagocytic Syndrome

The hemophagocytic syndrome is an uncommon complication of HIV infection that is characterised by proliferation of histiocytes and phagocytosis of marrow blood cell precursors. It typically presents with fever, pancytopenia, lymphadenopathy, and splenomegaly. The syndrome may be the result of infection with HIV itself or may occur in association with Epstein-Barr virus (EBV), CMV, *Herpes simplex* virus, tuberculosis, human herpesvirus 8 (HHV-8), or parvovirus B19 infections. Hemophagocytosis may also complicate malignancies like T cell non-Hodgkin's lymphoma or Kaposi's sarcoma.

Epidemiology of AIDS-Related Lymphoma

Lymphoma is a late manifestation of HIV infection, more likely to occur in the setting of significant immune suppression, with CD4 cells below 200/mm^3, CGH, and prior history of an AIDS-defining illness. Following an earlier diagnosis of AIDS, the relative risk of immunoblastic lymphoma is increased approximately 627-fold, whereas that of diffuse large cell lymphoma is 145-fold higher than that expected in the general population. Of interest, when linking cancer and AIDS registries, even low grade lymphoma was found to be increased 14-fold, while the incidence of T-cell lymphoma is also increased among patients with AIDS.

Highly active antiretroviral therapy (HAART) has resulted in a highly significant decline in mortality, and in development of new opportunistic infections, Kaposi's sarcoma, and primary central nervous system lymphoma among patients with AIDS., While significant decreases in systemic AIDS-related lymphoma have also occurred, the decline is not as profound as that seen in other AIDS-defining conditions, and lymphoma has now become one of the more common of the initial AIDS-defining illnesses.

Genetic Epidemiology of AIDS-related Lymphoma

AIDS-related lymphoma, similar to lymphoma occurring in immunocompetent hosts, is more common in men than in women. Although major environmental factors have not been associated with an increased risk for AIDS lymphoma, genetic factors in the host may be operative. Heterozygotes for the Ä32 deletion of the CCR5 coreceptor gene are statistically less likely to develop lymphoma, while those with SDF-1 mutations (32 A) are statistically more likely to develop lymphoma.

Cytokine Profile Associated with AIDS Lymphoma

The Multi-Centre AIDS Cohort Study (MACS) evaluated the cytokine profile in serum of homosexual/bisexual men who went on to develop lymphoma within the next 6-month period. Markers of activation, including soluble CD27, CD30, and CD44 were elevated when compared with HIV-positive controls who did not develop lymphoma. In addition, markers of isotype switching, including soluble CD23 and IgE, were elevated, as was serum interleukin (IL)-10, a marker of B-cell stimulation. Serum levels of immunoglobulin (Ig) M and IgG were decreased when compared with controls.

In multivariate analysis, serum CD23, CD27, IgM, IgG, IgE, and IL10 were independently associated with lymphoma. While it is highly likely that lymphoma had already developed in these patients, who were formally diagnosed within 1 to 6 months after serum collection, it is of interest that a particular pattern of markers, related to B-cell activation, stimulation and isotype switching were consistently expressed. These markers may provide the ability to predict which HIV-infected patients are at increased risk for development of B-cell lymphoma.

Standard versus Low-dose Chemotherapy

Prior to the development of HAART, phase II and III clinical trials demonstrated that low-dose chemotherapeutic regimens, such as mBACOD (methotrexate, bleomycin, doxorubicin, cyclophosphamide, vincristine, dexamethasone), were as efficacious as standard dose regimens, but had the advantage of statistically decreased rates of hematologic and other toxicity. With the addition of HAART to chemotherapy, standard dose therapy has become feasible and advantageous.

Use of Concomitant HAART Plus Chemotherapy

The use of HAART, along with dose-reduced and standard-dose CHOP (cyclophosphamide, doxorubicin, vincristine and prednisone) was evaluated by the National Cancer Institute (NCI)-sponsored AIDS Malignancy Consortium in a group of 65 patients with newly diagnosed AIDS-lymphoma. HAART therapy consisted of indinavir, stavudine, and lamivudine, the former of which is a protease inhibitor, while the latter 2 are nucleoside reverse transciptase inhibitors. Grade 3 or 4 neutropenia was more common among patients receiving full-dose CHOP (25% vs 12%), but there were similar numbers of patients with other toxicities.

Doxorubicin clearance and indinavir concentration curves were similar in patients on this study when compared to historical controls. Cyclophosphamide clearance was decreased 1.5-fold when compared to controls, although there was no apparent clinical consequence of this change. The authors concluded that HAART could be administered safely with concomitant low-dose or standard-dose CHOP chemotherapy. Caution should be used when using chemotherapy together with zidovudine, since this antiretroviral agent may cause significant bone marrow compromise in itself.

Infusional Cyclophosphamide, Doxorubicin and Etoposide

The 96 hour continuous infusion cyclophosphamide, doxorubicin, etoposide (CDE) regimen has been tested in a large, multi-institutional ECOG trial of 107 patients, including 48 who were also given antiretroviral therapy with didanosine (ddI), and 59 who received HAART regimens. For the group as a whole, the rate of complete remission was 44%, similar in patients who received either HAART or single agent didanosine. However, as has been described in the setting of other AIDS conditions, the median overall survival was significantly longer in AIDS lymphoma patients who received combination antiretroviral therapy. Thus, response rates to infusional CDE appear similar to those achieved with either low-dose or standard-dose m-BACOD, although survival is certainly superior in those patients who receive concomitant HAART.

EPOCH Regimen of Infusional Chemotherapy

Several biologic markers associated with prognosis in lymphoma were studied as part of this trial, and their impact upon prognosis was assessed. High MIB-1 activity (> 80%), associated with greater proliferative rate and poor prognosis in HIV-negative patients, was present in 85% of these HIV-positive individuals, but did not correlate with response or survival after EPOCH. Similarly, p53 expression, associated with poor prognosis in HIV-negative cases, was of no significance in terms of predicting survival in these EPOCH-treated patients. Bcl 2 expression, however, was predictive of shorter survival.

Most recent studies have demonstrated the advantage of combination antiretroviral and chemotherapy for AIDS-lymphoma, and there is no question that HAART has been associated with a remarkable prolongation in survival of these patients. Thus, Antinori and colleagues demonstrated that virologic response to HAART was the only factor associated with improved rates of complete remission to CHOP or CHOP-like chemotherapy. These data are in contrast to the NCI's results with EPOCH, in which outstanding rates of complete remission and long-term, disease-free survival were achieved, while purposely omitting HAART during chemotherapy. The immediate reinstitution of HAART at the completion of chemotherapy is clearly an important aspect of this approach.

Value of Rituximab when Combined with CHOP Chemotherapy in AIDS-Lymphoma

The addition of rituximab to the CHOP regimen has been associated with statistically significant improvements in response and survival among elderly patients with de novo diffuse large B-cell lymphoma. With these facts in mind, NCI sponsored the AIDS-Malignancy Consortium (NCI/AMC), which conducted a Phase III randomised trial of the standard CHOP regimen versus CHOP plus rituximab in a group of 151 patients with newly diagnosed AIDS-lymphoma.

The regimens employed included cyclophosphamide (750 mg/m^2, IV on day 1); doxorubicin (50 mg/m^2, IV on day 1), vincristine (1.4 mg/m^2 IV on day 1, capped at 2.0 mg total), and prednisone, 100 mg orally from days 1 through 5. In patients randomised to R-CHOP, rituximab was given at a dose of 375 mg/m^2 on day 1 of each cycle, while the same doses of CHOP were begun on day 3 of each 21day cycle. After attainment of CR, the R-CHOP patients also received 3 monthly doses of the antibody. Thus, a maximum of 9 to 11 doses of rituximab were given to R-CHOP treated patients.

All patients were treated until achievement of complete remission, plus 2 additional cycles, or a minimum of 6 cycles of therapy. Antiretroviral therapy, filgrastim, and prophylaxis for *Pneumocystis* pneumonia were mandated. Meningeal prophylaxis was not routinely given.

The results of this study are somewhat difficult to interpret, especially in terms of the reported increase in infectious deaths in the patients treated with R-CHOP. The fact that over 60% of these deaths occurred in patients with CD4 cells < 50/mm^3 would indicate the possibility that this fatal complication was more related to the severe underlying immune deficiency than to the use of rituximab per se.

The fact that there was no statistically significant difference in the incidence of infection, or in the occurrence of neutropenic sepsis would also mandate a consideration of the myriad factors that may predict whether a neutropenic patient with sepsis actually dies. Nonetheless, it may be important to note that a maximum of 9 to 11 doses of Rituximab were given to R-CHOP treated patients. This is substantially greater than that given in other settings of R-CHOP, and may have contributed to the findings.

The finding that R-CHOP was no more efficacious than CHOP alone in the AMC/NCI trial is also noteworthy. The R+ CHOP treated patients tended to have a higher CR rate than those treated with CHOP alone. Further, 19.5% of CHOP treated patients died of progressive lymphoma, versus 10% of those who received CHOP with rituximab. A recent Phase II French study of R-CHOP in patients with AIDS-lymphoma reported a higher complete remission rate (77%) than that reported by the AMC/NCI (59%). Nonetheless, the majority (54%) of the French patients had low-risk IPI scores (0 or 1), and the study was therefore "driven" by the inclusion of these individuals.

Further, the median CD4 cell count was relatively high (180/mm^3), and the median HIV viral load was relatively low (9236 copies/cc) in the French study. It is possible that the excellent results in the French R+ CHOP trial were due to the addition of rituximab to CHOP.

Alternatively, these results may also be due to the fact that a relatively low-risk group of patients was enrolled. The French study was also smaller than that of the NCI/AMC, and was not randomised.

It is clearly possible that rituximab does *not* improve the response rate of patients with AIDS-lymphoma, when added to the standard CHOP regimen. In this regard, Bcl-2 has been associated with poor prognosis in patients with aggressive B-cell lymphoma. Bcl-2 expression is also important in determining response to R-CHOP in elderly HIV negative patients, with Bcl-2 positive cases more likely to benefit from addition of rituximab.

It is certainly possible that AIDS related aggressive lymphomas may be less likely to express Bcl-2, and thus less likely to benefit from the addition of rituximab. In this regard, Little and colleagues from the NCI have compared Bcl-2 expression in 25 HIV infected and 33 HIV negative patients with diffuse large B-cell lymphoma.

Bcl-2 was expressed in 16% of the AIDS related cases, and 41% of the *de novo* lymphomas. If these data hold true in larger series, one might expect that addition of rituximab would be less beneficial in the setting of AIDS-lymphoma, simply because Bcl-2 is less frequently expressed.

A full understanding of the results of the current AMC/NCI trial will require further elucidation of Bcl-2 status of these patients, and further detailed analysis of the factors associated with infectious death among the R-CHOP treated patients.

Use of CNS Prophylaxis in Patients with Systemic AIDS-lymphoma

No prospective, randomised study has yet addressed the issue of CNS prophylaxis in patients with AIDS-lymphoma. However, the risk of leptomeningeal (cerebrospinal [CSF]) relapse is a real one in this setting, reported in approximately 15–20% of patients, and more likely to occur in patients with bone marrow involvement.

The majority of trials in the US have included intrathecal chemotherapy (methotrexate or cytosine arabinoside), given a total of 4 to 6 times during induction chemotherapy. In the recent NCI trial of dose adjusted infusional EPOCH, CNS prophylaxis was added to the last 17 patients, after late CSF relapse occurred in earlier patients who did not receive such therapy.

Involvement of the CNS is an independent predictor of poor prognosis in patients with systemic AIDS-lymphoma.

Therapy of Patients with Relapsed or Primary Resistant AIDS-lymphoma

Treatment options for patients with relapsed or refractory AIDS-lymphoma are limited. The infusional CDE regimen has been associated with a complete remission rate of 4% in a group of 24 patients with relapsed/refractory disease, and a median survival of 2 months. A regimen consisting of etoposide, prednimustine, and mitoxantrone resulted in complete response in 8 (38%) of 21 patients but a median survival of only 2 months.

While associated with uniform grade 4 neutropenia, the ESHAP regimen, when given to 13 patients with relapsed or refractory AIDS-lymphoma, led to complete remission in 31%, overall response in 54%, and median survival of 7.1 months from the time of ESHAP. High-dose chemotherapy, followed by autologous stem cell rescue has been efficacious in patients with relapsed/refractory AIDS-lymphoma, with complete remission rates and toxicity profiles similar to that seen in HIV negative patients, provided that effective antiretroviral therapy is also employed.

Primary Effusion Lymphomas

Primary effusion lymphoma (PEL) is an aggressive B-cell lymphoma, usually confined to body cavities, which occurs primarily, but not exclusively, in HIV-infected patients. Patients with HIV-associated PEL tend to be homosexual/bisexual males, with severe immunosuppression. PEL is caused by a gamma 2 herpesvirus termed Kaposi sarcoma-associated herpesvirus (KSHV), also known as human herpesvirus 8 (HHV-8). HHV-8 is also the cause of Kaposi sarcoma, from which it was first isolated.

The malignant cells in PEL are large and pleomorphic, and may resemble Reed Sternberg cells, although they are CD15 negative. Phenotypically, the cells stain for leukocyte common antigen CD45 and various activation antigens (HLA-DR, EMA, CD30, CD38, CD77), but are usually negative for other B- and T-cell markers including CD20 and CD19. The B-cell nature can be demonstrated by the presence of immunoglobulin gene rearrangements by Southern blot or PCR.

The malignant cells in PEL are derived from postfollicular B cells, which harbour somatic hypermutations of the immunoglobulin genes. The pathogenesis of PEL is of interest, since HHV-8 carries a number of oncogenic sequences including a Bcl-2 like sequence, a G-coupled receptor, and a type D cyclin similar to PRAD1. The virus also produces cytokines such as viral IL-6, capable of contributing to angiogenesis and tumour cell growth.

Patients with PEL classically present with symptoms and signs of an effusion (pleural, pericardial, or ascitic) in the absence of a tumour mass. Extension to adjacent structures such as the chest wall, pleura, or peritoneum may also occur. Rarely, PEL may be diagnosed in the absence of an effusion, either at diagnosis or during the course of disease.

With standard CHOP chemotherapy, median survival has been in the range of only 60 days. Several case reports have documented the successful use of HAART, either with or without antiherpetic therapy. While the PEL cells do not classically express CD20, a case of pathologically confirmed complete remission of PEL after HAART and rituxan has also been described.

Castleman's Disease in the Setting of HIV Infection

Multicentric Castleman's disease (MCD) is a polyclonal lymphoproliferative disorder characterised by recurrent fevers, lymphadenopathy, hepatosplenomegaly, and autoimmune phenomena, which often progresses to malignant lymphoma. The plasma cell variant of MCD has been described in the setting of underlying HIV infection, and appears to be associated with HHV-8 infection, with a high degree of HHV-8 lytic gene activity and high-titer viremia. While cidofovir and ganciclovir have been effective against HHV-8 in vitro, cidofovir was shown to be ineffective when used in patients with AIDS-KS.

Nonetheless, AIDS patients receiving ganciclovir for cytomegalovirus retinitis had a 40% reduction in the risk of developing KS over time, suggesting that antiherpetic therapy may be of some clinical value in HHV-8 related diseases. The optimal therapy of HIV associated MCD is unknown. Recently, Casper and colleagues demonstrated the efficacy of ganciclovir (5 mg/kg twice daily for 1 week, and then once daily) given to 3 HIV infected patients with MCD. Rituximab has also shown activity in this setting, with complete remission attained in 4 of 6 such patients.

Of interest, splenectomy has also been efficacious in HIV-infected patients with MCD, associated with resolution of fevers, and improvement in anemia. Long-lasting remission was described in 8 of 10 such patients, after receipt of both splenectomy and HAART.

Hodgkin's Disease in the Setting of HIV Infection

While not considered an AIDS-defining illness, the incidence of Hodgkin's disease (HD) is clearly increased among HIV-infected individuals. Unusual clinical and pathologic characteristics of HD have been described in this setting. Thus, systemic "B" symptoms are almost always present, mixed cellularity HD is the predominant pathologic subtype of disease, and advanced, extranodal disease is expected in the majority.

Bone marrow involvement has been documented in 40% to 60% of patients at initial diagnosis, and patients often undergo the initial diagnostic bone marrow examination for the evaluation of fever of unknown origin in the setting of HIV infection and pancytopenia. While standard multiagent chemotherapy may be curative in most HIV negative patients with stage III or IV HD, the median survival for HIV infected patients has been in the range of 1 to 2 years. Standard-dose ABVD with hematopoietic growth factor support was evaluated in a multi-institutional trial of 21 HIV infected patients. Antiretroviral therapy was not used. Neutropenia (< 500 cells/mm^3) developed in almost 50%, and median survival for the group was only 18 months.

It is possible that results would have improved with concomitant use of highly active antiretroviral therapy (HAART), as was demonstrated with the Stanford V regimen, employed in 59 HIV infected patients from Italy. In this study, complete remission was attained in 81%, and 56% of these (i.e., 45% of all patients treated) are estimated to remain disease-free at a median follow-up interval of 33 months. Grade 3 or 4 neutropenia occurred in 78%, despite use of G-CSF. Further work will be required to define the optimal therapy for such patients.

Stem Cell Mobilisation

Successful blood and marrow transplant (BMT), both autologous and allogeneic, requires the infusion of a sufficient number of hematopoietic progenitor/stem cells (HPCs) capable of homing to the marrow cavity and regenerating a full array of hematopoietic cell lineages in a timely fashion. At present, the most commonly used surrogate marker for HPCs is the cell surface marker CD34, identified in the clinical laboratory by flow cytometry. Clinical studies have shown that infusion of at least 2×10^6 CD34$^+$ cells/kg recipient body weight results in reliable engraftment as measured by recovery of adequate neutrophil and platelet counts approximately 14 days after transplant. Recruitment of HPCs from the marrow into the blood is termed mobilisation, or, more commonly, stem cell mobilisation.

In Section I, Dr. Tsvee Lapidot and colleagues review the wide range of factors influencing stem cell mobilisation. Our current understanding focuses on chemokines, proteolytic enzymes, adhesion molecules, cytokines and stromal cell-stem cell interactions. On the basis of this understanding, new approaches to mobilisation have been designed and are now starting to undergo clinical testing.

In Section II, Dr. Michele Cottler-Fox describes factors predicting the ability to mobilise the older patient with myeloma. In addition, clinical approaches to improving collection by individualising the timing of apheresis and adjusting the volume of blood processed to achieve a desired product are discussed. Key to this process is the daily enumeration of blood $CD34^+$ cells. Newer methods of enumerating and mobilising autologous blood HPCs are discussed.

In Section III, Dr. John DiPersio and colleagues provide data on clinical results of mobilising allogeneic donors with G-CSF, GM-CSF and the combination of both as relates to the number and type of cells collected by apheresis. Newer methods of stem cell mobilisation as well as the relationship of graft composition on immune reconstitution and GVHD are discussed.

Current Understanding of Factors Influencing Stem Cell Mobilisation

Recruitment of hematopoietic stem and progenitor cells to the blood following treatment with chemotherapy, or cytokines, is a clinical process termed mobilisation. This process mimics enhancement of the physiological release of stem cells and progenitors from the bone marrow (BM) reservoir in response to stress signals during injury and inflammation. Currently, mobilised cells are the preferable and major source of stem and progenitor cells harvested for autologous and allogeneic transplantations because of the higher yield of these cells, leading to faster engraftment and decreased procedural risks compared with harvested BM cells.

The emerging picture of stem cell mobilisation involves interfering with the physiological interplay between mesenchymal stromal and hematopoietic cells regulating both bone and BM remodelling processes, which also mediate stem/progenitor cell proliferation and migration. The mobilisation process is initiated by stress-induced activation of neutrophils and osteoclasts by chemotherapy and repeated stimulation with cytokines such as granulocyte colony-stimulating factor (G-CSF), resulting in shedding and release of membrane-bound stem cell factor (SCF), proliferation of progenitor cells, as well as activation and/or degradation of adhesion molecules such as VLA-4 and P/E selectins.

The dynamic secretion and inactivation of the chemokines stromal cell derived factor-1 (SDF-1)/CXCL12 and interleukin-8 (IL-8)/CXCL8, and multiple cycles of inactivation and degradation of BM components by proteolytic enzymes, such as elastase, cathepsin G, proteinase 3, CD26, and various matrix metalloproteinases (MMPs), are implicated as major players in progenitor mobilisation. These mechanisms will be reviewed in this section.

The road to stem cell mobilisation began in the 1960s, with publication of a few reports documenting the presence of hematopoietic stem cells in the peripheral blood of mice, dogs, and monkeys, followed by reports revealing low levels of progenitors in the human circulation during steady-state homeostasis. However, steady-state progenitors in the circulation of mice were shown to be inferior to BM as a source of repopulating stem cells. More important, although sporadic clinical transplantations documented successful recovery, in most cases graft failure was detected in autologous, syngeneic, and fully matched related allogeneic transplantations, using steady-state peripheral blood cells.

Moreover, the first successful human blood leukocyte transplantations were not better than BM as a source of repopulating stem cells (reviewed in Korbling and Fliender and To et al). While the levels of human progenitors in the circulation during steady-state homeostasis are low, they are significantly elevated in patients with myeloproliferative disorders. Similarly, a transient increase in circulating progenitors was documented in dogs treated with dextran sulfate and humans treated with endotoxin or other stress-inducing mobilisation agents.

These encouraging results were followed by preliminary reports in the late 1970s documenting increased levels of progenitors in the circulation of patients after chemotherapy treatment with cyclophosphamide and other drugs. Technical improvements including in vitro colony assays and the availability of large-scale harvesting of human progenitors by continuous-flow leukopheresis enabled closer examination of chemotherapy-/druginduced mobilisation in treated patients. In the early 1980s, To and colleagues reported high levels of progenitors in the circulation of leukemia patients in early remission. Clinical studies by this group were the first to document the beneficial faster repopulation in patients transplanted with autologous mobilised peripheral blood (PBL). Initially, the mobilisation protocols were based on chemotherapy alone. However, after the discovery of human G-CSF by Welte et al, mobilisation protocols began to include G-CSF, which today is the standard mobilising agent.

Definitive, Repopulating Stem Cells are Migrating Cells

During late embryonic development, both human and murine stem cells migrate via the blood circulation from the fetal liver to the BM and repopulate this tissue with high levels of immature and maturing cells. These in turn are released into the circulation, while a small pool of undifferentiated stem and progenitor cells within the BM is maintained throughout life. Prior to their localisation in the murine fetal liver, hematopoietic stem cells can be isolated from the aorta-gonad-mesonephros (AGM) region and the yolk sac even before completion of blood circulation development.

However, some of these stem cells are not definitive, since they cannot rescue and repopulate lethally irradiated recipients because of their inability to migrate in the host circulation and home to and repopulate the BM. These predefinitive stem cells isolated from the murine yolk sac can still engraft the liver of newborn mice preconditioned as fetuses with chemotherapy if the predefinitive stem cells are directly injected into this organ. Furthermore, an additional maturation step provided to predefinitive progenitors isolated from the AGM region, in the form of in vitro incubation on stromal cells and stimulation with cytokines, shifts these cells into definitive stem cells with functional migration, homing, and repopulation potential.

Murine embryos that lack the chemokine SDF-1 or its receptor CXCR4 have multiple lethal defects, including impaired seeding of the fetal BM (reviewed in Lapidot and Petit). We have shown the essential roles of SDF-1/CXCR4 interactions in a functional, preclinical model for human stem cells using non-obese diabetes severe combined immune deficient (NOD/SCID) mice as recipients. Homing and repopulation in both primary and serially transplanted mice with human CD34 stem cells was dependent on CXCR4 signalling, which is dynamically expressed by human progenitors.

Moreover, SDF-1 mediated migration of human CD34 cells in vitro correlated with their in vivo repopulation potential in transplanted patients. In conclusion, stem cell motility (migration, homing, and release) is essential for BM repopulation and for the development of the hematopoietic system.

Hematopoietic Stem Cells in the Circulation Can Home Back to the BM

The role of circulating stem cells in steady-state homeostasis is currently unknown. While some progenitors are needed for seeding of the thymus, which requires migration of lymphocyte precursors from the BM, this process is markedly reduced in adult life. Stem cells migrate also to nonhematopoietic organs such as the liver, especially during liver injury/inflammation, creating an alarm situation and transmitting stress signals that mobilise and recruit stem cells as part of organ repair. Ramshaw et al suggest that circulating stem cells can reengraft the BM, documenting successful engraftment of unconditioned murine recipients transplanted with high doses of BM cells.

In support of this approach, Wright et al elegantly demonstrated rapid clearance of intravenous transplanted mouse stem cells from the murine circulation within a few minutes. By using genetically marked parabiotic mice with a shared circulation, they showed that the low levels of circulating stem cells can durably reengraft the BM, suggesting continuous release and reengraftment of the BM as sequential events taking part in physiological processes. In parallel, Abkowitz et al, using the same parabiotic model, revealed only low levels of reengrafted stem cells with secondary recipient repopulation potential (1%–2.5%) in the partner BM as opposed to high levels of progenitors in the spleen.

These investigators suggest the release of stem cells into the circulation functions also as an apoptotic pathway for the BM to regulate excess amounts of progenitors because of limited numbers of stem cell niches. Of interest, these investigators also applied G-CSF- and SCF-induced mobilisation to the parabiotic mice pairs, documenting a dramatic 4 to 10-fold increase (from 1%–2.5% to 10.1%) in the levels of reengrafted stem cells with secondary repopulation potential in the partner BM.

These results demonstrate that inducing the release of stem cells in response to stress signals also increases their homing back and reengraftment of the stimulated/stressed BM, confirming the repopulation potential of mobilised progenitors harvested for clinical protocols. These results also support the hypothesis that tissue repair of damaged BM can be supported by intensified homing and reengraftment of circulating hematopoietic stem cells, in response to stress signals, which is similar to migration of mature leukocytes to sites of inflammation as part of the immune system host defence. The concept of BM niches occupied by stationary, quiescent stem cells may be an oversimplification of a more dynamic situation.

Higher proliferative and migratory turnover of definitive, repopulating stem cells involves their release into the circulation, migration, homing, and reengraftment of the BM as part of their physiological development. Bradford et al revealed that even the most primitive murine stem cells in the BM must cycle at least once every 30 days.

Bone and BM Remodelling Processes are Mutually Regulated

Proliferation and release of hematopoietic stem and progenitor cells require dynamic cycles of BM destruction/restructuring, which seem to be linked to bone remodelling by osteoclast/osteoblast interactions. Moreover, both processes are likely to be orchestrated by the same factors. For example, continuous G-CSF treatment to prevent neutropenia has also been shown to induce osteoporosis in some juvenile patients with severe congenital neutropenia and in transgenic mice that overexpress this cytokine.

More important, there is a dramatic increase in the number of murine osteoclasts in response to repetitive G-CSF stimulation, leading to osteoclast-mediated bone resorption during stem cell mobilisation and calcium release, which can induce detachment of hematopoietic progenitors from fibronectin in vitro. Unexpectedly, treatment of mice with pamidronate, an inhibitor of osteoclast-mediated bone resorption, together with G-CSF increased the levels of mobilised progenitors in the murine circulation. Osteoclasts secrete high levels of IL-8, a mobilisation-inducing chemokine.

Another chemokine, SDF-1, which also participates in G-CSF mediated mobilisation, induces osteoclasts to secrete the metalloproteinase MMP-9. This enzyme is involved in stem cell mobilisation by shedding the membrane-bound cytokine SCF within the BM and releasing this cytokine into the circulation. MMP-9 also induces migration of osteoclasts to the cortical bone, adjacent to the SDF-1 and stem cell–rich endosteum region, to form resorption pits. These results suggest that the role of osteoclasts in mobilisation could be related to the rapid increase in their numbers and to the secretion of chemokines and proteolytic enzymes. Activation and secretion of proteolytic enzymes lead to cleavage and degradation of the BM extracellular matrix, adhesion molecules, cytokines, and chemokines and therefore facilitate transendothelial migration and stem cell release from the BM. These processes mediate both peripheral bone and medullar BM remodelling as well as stem cell proliferation and release by inactivating the BM microenvironment.

Subsequently, stem cells migrate via the circulation, home back to the BM, and repopulate the damaged/destructured sites in this organ as part of tissue repair and the continuous replenishment of the blood with new immature and maturing hematopoietic cells of all lineages, while maintaining a small pool of undifferentiated stem and progenitor cells within their renewed BM niches. In support of this hypothesis, by applying complementary DNA array technology, 2 recent studies revealed high expression of the proteolytic enzyme proteinase 3 and low expression of a proteinase 3 inhibitor in isolated human BM CD34 cells compared with enriched mobilised CD34 cells isolated from G-CSF treated healthy donors. These results suggest that proteinase 3, with its ability to cleave connective tissue in the BM, is needed for maintaining a niche for hematopoietic stem and progenitor cells.

Recent findings document that proteinase 3 also cleaves the cell cycle checkpoint p21, inducing cells to enter the cell cycle. This could explain the higher levels of cycling BM CD34 cells compared with the more quiescent mobilised PBL CD34 cells. Immature mesenchymal bone-forming osteoblasts lining the stem cell–rich endosteum region secrete high levels of SDF-1. In response to stimulation with this ligand, hematopoietic stem cell–derived osteoclasts secrete MMP-9. Thus, peripheral periosteal bone and medullar BM destruction and remodelling are sequential events mediated by the same factors, i.e. MMP-9, G-CSF, SCF, IL-8, and SDF-1, which regulate both new bone formation and stem cell self-renewal, migration, and development.

Stress-Induced Mobilisation of Stem and Progenitor Cells

Applying single or multiple stress signals interferes with steady-state homeostasis, creating an alarm situation leading to increased release of immature and maturing hematopoietic cells from the BM into the blood circulation, which occurs naturally during inflammation and injury. For example, increased levels of G-CSF and other cytokines secreted by long-distance runners during marathons cause blood leukocytosis, in particular higher levels of CD34/CD38-positive committed progenitor cells, but not more primitive CD38-negative stem cells compared with control individuals, reflecting adaptation of

BM activity to stress induced by extensive exercise. Increased cell release from the BM reservoir is part of the immune system host defence during inflammation as a result of infection- or injury-mediated release of stress signals.

This release is induced clinically or in animal models by a wide range of molecules and/or treatments: DNA damaging, single chemotherapeutic drugs such as cyclophosphamide (Cy), or combined chemotherapy regimens such as iphosphamide, carboplatin and etoposide (ICE) and etoposide, methylprednisolone, ara-c and cisplatin (ESHAP); cytokines such as G-CSF, granulocyte-macrophage colony-stimulating factor (GM-CSF), SCF, and flt-3 ligand; and chemokines such as IL-8, Mip-1a, Groâ, and SDF-1. These molecules differ in their mode of administration, the time frame needed to achieve mobilisation, the type of cells mobilised, and the efficiency. G-CSF, the most commonly used mobilisation agent, is usually administered daily at a dose of 5–10 ig/kg for 5–10 days, alone or after chemotherapy.

However, a substantial number of patients, especially patients having extensive chemotherapy that leads to BM aplasia, older patients, and a minority of healthy individuals, are poor mobilisers. For example, increasing age and longer intensive chemotherapy treatment in old multiple myeloma patients inversely correlate with successful mobilisation. These patients are usually treated with higher doses of G-CSF, GM-CSF followed by G-CSF, or a combination of G-CSF and SCF. Recently, it was reported that some chronic myeloid leukemia (CML) patients treated with high doses of imatinib (Glivec, STI 571) suffer hematological toxicity and can develop drug-induced neutropenia.

In addition, some treated patients failed to mobilise well, requiring discontinuation of the drug during the mobilisation process and the addition of SCF to G-CSF, to successfully overcome the poor mobilisation. However, while SCF stimulation increases the levels of mobilised CD34 progenitors, this treatment also has side effects such as allergic reaction due to activation of mast cells.

These studies stress the need to identify and characterise the mechanisms of stem cell mobilisation in order to develop better strategies for poor mobilisers. Chemotherapy followed by repetitive G-CSF induced mobilisation is a multistep process. Anchorage of stem and progenitor cells to the BM microenvironment by activation of adhesion interactions, which are facilitated by membrane-bound SCF or SDF-1, needs to be disrupted. Papayannopoulou elegantly demonstrated that the integrin VLA-4 plays a critical role in stem cell mobilisation since anti-VLA-4 antibodies prevent adhesion molecules expressed on the surface of hematopoietic progenitors from binding their ligand VCAM-1 expressed by stromal cells. This induces rapid mobilisation of primate and murine stem cells, which also involves signalling via the c-kit receptor. Moreover, as a result of inactivation during the mobilisation process, mobilised CD34 cells have lower levels of VLA-4 and c-kit expression compared with BM progenitors.

In addition, sulfated polysaccharides such as Fucoidan lead to rapid mobilisation of progenitors since they compete for adhesion interactions with selectins and for SDF-1 binding to the BM endothelium via its nonsignalling C-terminus, inducing inactivation of selectin interactions as well as release of SDF-1 into the circulation. Levesque et al revealed increased release of elastase and cathepsin G within murine BM during Cy- and G-CSF induced mobilisation; peak levels were reached during stem cell egress. These enzymes cleaved VCAM-1 expressed by stromal cells, preventing adhesion of progenitors via VLA-4 as part of the mobilisation process.

In addition to their distinctive adhesive properties, mobilised CD34 progenitors differ from their BM counterparts in several ways. Besides reduced VLA-4, c-kit, and CXCR4 expression, a significantly higher percentage of mobilised progenitors are noncycling quiescent cells. Prior to mobilisation, hematopoietic progenitors proliferate within the BM, and this feature could involve proteinase 3–mediated cleavage of p21 in response to neutrophil stimulation.

Mobilised progenitors have higher levels of the proapoptotic genes caspase 3, 4, and 8 and reduction in inhibitors of apoptosis such as antiproteinase 2, compared with human BM CD34 cells, supporting the hypothesis of Abkowitz et al that release into the circulation may also serve as an apoptotic pathway for stem cells. Enhanced repopulation documented with mobilised progenitors is due to significantly higher cell doses, in particular committed progenitors, which lead to faster neutrophil and platelet recovery. However, better leukemia-free survival and overall survival were recently documented in acute myeloid leukemia (AML) patients transplanted with increased cell doses of matched BM cells compared with lower doses of BM cells and more important, also compared with high cell doses of matched mobilised PBL, suggesting BM as a superior source of stem cells for HLA-identical allogeneic transplantations.

New protocols aimed at increasing the levels of long-term repopulating stem cells as the major source while maintaining short-term repopulating cells need to be developed.

SDF-1/CXCR4—Key Regulators of Stem Cell Homing and Mobilisation

Mobilised human CD34 progenitors express reduced levels of the SDF-1 receptor CXCR4, which correlates with improved mobilisation, suggesting involvement of SDF-1/CXCR4 interactions in the mobilisation process. Overexpression of SDF-1 in the murine circulation leads to stem cell mobilisation. In functional preclinical animal models for human stem cells such as preimmune sheep and immune-deficient NOD/SCID mice, mobilised CD34 cells are inferior in their repopulating potential compared with similar cell doses of CD34 cells obtained from human BM.

This is most probably due to their reduced levels of surface CXCR4, which is needed for SDF-1 mediated directional homing and repopulation in transplanted mice and increased levels of short-term repopulating cells. These results should be interpreted with caution, since short-term stimulation with human SCF upregulated surface CXCR4 expression on human CD34 cells and murine SCF is much less potent in stimulating human progenitors. In addition, other factors such as complement C3a improve SDF-1 mediated directional migration of human progenitors and the in vivo homing of murine progenitors. However, some of these factors can be species-specific or absent in immune-deficient mice.

Of interest, priming of mobilised human progenitors by several inflammatory molecules improves their chemotactic responses to SDF-1. We have demonstrated increased levels of SDF-1 production in the murine BM in response to DNA-damaging chemotherapy drugs, including Cy and 5-fluorouracil (5-FU), most probably to prevent cell death since this chemokine is also a survival factor for stem cells. Rafii and Hattori et al demonstrated that the increased levels of SDF-1 in response to stress-mediated 5-FU mobilisation in mice induce release of MMP-9, which cleaves and releases membrane-bound SCF, leading quiescent stem cells to proliferate in the BM prior to their mobilisation into the circulation.

SDF-1 mediates secretion of MMP-2 and MMP-9 from human CD34 cells, and these proteolytic enzymes inactivate SDF-1 by cleaving a few amino acids in the N terminus. We demonstrated that each injection of G-CSF stimulates mesenchymal cells such as immature osteoblasts to secrete SDF-1, leading to a transient increase in SDF-1 levels within the BM. This oscillating increase is followed by a profound decrease due to inactivation by proteolytic enzymes, mostly by neutrophil elastase, with the lowest levels of this ligand released during cell mobilisation. Levesque et al reported that in addition to inactivation of BM SDF-1 by elastase and cathepsin G, these proteolytic enzymes can also cleave part of the CXCR4 receptor N terminus in the BM, partially inactivating SDF-1 signalling and directional migration.

Of interest, neutrophils secrete the proteolytic enzymes elastase, cathepsin G, and proteinase 3 in response to G-CSF, and neutrophil elastase also inactivates G-CSF, which is essential for regulating release of neutrophils from the BM. However, expression of G-CSF receptors by neutrophils is not required because of an indirect mechanism that involves proteolytic reduction of SDF-1 within the murine BM. Of interest, the central role of elastase in regulating release of BM leukocytes is suggested by inherited genetic mutations in the elastase gene, leading to 2 forms of neutropenia in juvenile patients (severe congenital neutropenia, or Kostmann disease, and cyclic neutropenia) that require continuous treatment with G-CSF. CD26, another proteolytic enzyme that inactivates SDF-1, is expressed by human CD34 progenitors and is involved in G-CSF induced mobilisation in mice.

A role for CXCR4 signalling in cell egress from the BM to the circulation emerges from several studies. We found that in parallel to gradual reduction of BM SDF-1 during G-SCF administration, CXCR4 expression is elevated within the human and murine BM, reaching peak levels at the time of mobilisation. While some CXCR4 inhibitors also lead to release of human progenitors, and treatment with pertussis toxin, which blocks Gai-mediated signalling in 7 transmembrane G coupled receptors such as CXCR4, induced mobilisation in mice, SDF-1/CXCR4 interactions are also needed for cell egress.

Treatment of mice with G-CSF and neutralising anti-CXCR4, anti-SDF-1 Ab, or inhibitors of CD26 reduced the mobilisation levels. Finally, warts, hypogammaglobulinemia, immunodeficiency and myelokathexis (WHIM) syndrome, a rare inherited immunodeficiency disease associated with mutations in CXCR4, is characterised by neutropenia and B-cell lymphopenia. In one case, an affected individual was born with cardiac malformation, resembling the murine model in which knocking out CXCR4 or SDF-1 results in a lethal phenotype associated with lack of BM seeding, B-cell development, and cardiac septum formation. Thus, SDF-1/CXCR4 interactions may have a role in the regulation of the routine and active egress of progenitor and maturing cells from the BM into the circulation.

Taken together, these results decipher key mechanistic insights into stress-induced mobilisation that mimic and amplify naturally occurring recruitment of progenitors during alarm situations. We suggest the BM as a reservoir for immature and maturing hematopoietic cells to be released into the circulation upon stress signals, to migrate to injured sites, and to contribute to host defence and tissue repair. The steady-state balance in the BM is disrupted, leading to transient increased production of SDF-1, and proliferation and activation of neutrophils and osteoclasts.

Release of proteolytic enzymes is followed by shedding of membrane-bound SCF, proliferation of hematopoietic progenitors, increasing surface CXCR4 expression and inactivation of SDF-1, G-CSF, the BM adhesion machinery, and extra cellular matrix (ECM). These events are intensified in each cycle of stimulation by G-CSF, eventually leading to release of progenitors into the circulation contains a model of the process. A better understanding of the process by which progenitors egress from the BM will eventually lead to the development of improved mobilisation protocols, in particular for patients who are poor mobilisers.

Because of the complexity of stem cell mobilisation and the wide range of agents inducing this process, we could not discuss many interesting studies, and we apologize to those whose excellent work could not be cited given the space limitations.

Enumeration of HPCs

Variables having an impact on the ability to collect hematopoietic progenitor cells (HPCs) may be donor related or procedure related. Donor-related variables include age, previous chemotherapy, mobilisation regimen, and platelet count at the time of mobilisation. Procedure-related variables include central access devices as well as variables inherent to the different cell separation devices used.

Nonetheless, at present, the basis for optimising HPC collection is the ability to determine when to start collection, based on the ability to enumerate $CD34^+$ cells in the peripheral blood of a donor/patient on a daily basis. It has been demonstrated repeatedly that the best currently available predictor of an adequate collection is the number of CD34$^+$ cell/iL in the blood on the morning of collection, both for good mobilisers and for poor mobilisers., A number of authors have recommended starting collection when a particular number of CD34$^+$ cells/iL is present (usually a number between 8 and 20) in order to increase the likelihood of collecting at least 2–4 × 10^6 CD34$^+$ cells/kg in a single apheresis, i.e., an acceptable number of HPCs for either one or two autologous transplants or a single allogeneic transplant.

In the past it was important that each centre determine a number of its own as the starting point for apheresis, given the historic difficulty in comparing flow cytometry results from centre to centre. At present, however, there are two commercially available single platform tests available for measuring CD34$^+$ cells in blood (ProCount from Becton-Dickinson, Mt View, CA, and StemKit from Beckman-Coulter, Fullerton, CA), so that any centre using the same technology should have a reasonable chance of success using the same numbers.

More importantly, it is possible to use the number of CD34$^+$ cells/iL obtained using single platform technology in a predictive formula which allows the clinician to know when to start apheresis in order to optimise collection and how many litres of blood need to be processed in order to collect a given number of CD34$^+$ cells using the formula in While results of collection using this formula have only been reported for the COBE Spectra cell separators, it seems likely that it would work with any continuous flow apheresis device if the appropriate value for machine collection efficiency is used.

Predictive Formula that Allows the Clinician to know when to Start Apheresis in Order to Optimise Collection and How many litres of Blood Need to be Processed in Order to Collect a given Number of CD34$^+$ Cells

Sadly, despite the consensus that daily CD34$^+$ cell numbers in the blood are the best currently available predictor for achieving an acceptable HPC collection, the practice has not been universally adopted because it is both time consuming and relatively expensive. Alternatives that have been reported to be useful include measuring the number of CD34$^+$ cells in the blood the day before starting collection, with or without the total white cell count, or change in white cell count from the day before apheresis to the day of apheresis, and the rapidity of rise in white cell count and platelet count.

More recently, the Food and Drug Administration (FDA) has approved the use of an HPC window on an automated cell counter (Sysmex, Kobe, Japan) and this number may be used to predict when to start apheresis. The HPC number measured by the Sysmex does not correlate well with CD34$^+$ cell number in the blood (nor does CD133, an antigen expressed by the more primitive CD34$^+$ cells [Cottler-Fox et al, in preparation]), but since it is now recognised that not all HPCs express measurable CD34 it has been presumed that the HPCs measured by the Sysmex include some that are CD34$^-$. Thus, the HPCs measured by Sysmex are available rapidly and less expensively, and may serve as a guide for starting apheresis, but they are difficult to use in the above predictive formula.

HPCs expressing high levels of the enzyme aldehyde dehydrogenase (ALDH) are believed to be pluripotent and generally express CD34 on their surface. A commercial assay has now been developed for this intracellular enzyme (Aldecount, Stemco Biomedical, Durham, NC). As this assay detects both CD34$^+$ and CD34$^-$ cells, and appears to differentiate viable from non-viable cells, its use may eventually lead to a major change in how HPCs are enumerated in transplant grafts.

Timing of Apheresis

There is a fall in circulating leukocytes immediately after granulocyte colony-stimulating factor (G-CSF) has been given, with the peak of CD34$^+$ cell mobilisation approximately 3–6 hours after each dose is given subcutaneously, paralleling its half-life., To optimise collection, it seems reasonable therefore, to wait at least 1 hour after giving the injection, trying to time the duration of the collection for the period which includes the rise in leukocytes and as much of the peak as possible. For example, at least one centre routinely waits for 2 hours after the injection, since their average collection lasts about 2–3 hours.

It is possible that the apparent ability of large volume leukapheresis lasting 4–5 hours to "recruit" CD34$^+$ cells into the blood, reported by several groups but not others, in fact simply reflects the timing of the apheresis collection relative to the dose of G-CSF. With the arrival of pegylated filgrastim (Neulasta, Amgen, Thousand Oaks, CA) it is possible that the relationship of collection yields relative to the time of dosage will cease to be a factor in collection.

Predicting Who Will Mobilise Poorly

The effect of age has been shown to be a continuous, incremental variable in the myeloma population, with no threshold past which there is an accelerated decline in mobilisation of CD34$^+$ cells. That is, the older the patient, the fewer stem cells are likely to be collected, but there is no fixed age past which it is impossible to collect HPCs. Over and above age, however, two other variables are statistically significant: number of months of previous chemotherapy and platelet count at the time of mobilisation. In 85% of myeloma patients over age 70 who had < 12 months of therapy and a platelet count > 200 × 10^9/L, it was possible to obtain e" 4 × 10^6 $CD34^+$ cells/kg in a single apheresis using individually optimised collection conditions after combined chemotherapy and growth factors. Those patient over 70 years of age with > 12 months of prior therapy and platelets < 200 × 10^9/L, however, were poor mobilisers.

Choosing a Mobilisation Regimen

Mobilisation with chemotherapy and growth factors has been shown in a number of settings to be significantly more effective than growth factors alone. Nonetheless, it is often the case that the toxicity of chemotherapy makes this combination less attractive in the fragile older patient. It is therefore important to know that in the subgroup of myeloma patients over the age of 70 who were predicted to be poor mobilisers (> 12 months prior therapy and platelets < 200 × 10^9/L), as many stem cells were collected with growth factors alone as with chemotherapy and growth factors together. Thus, for this group of patients it seems worthwhile trying to mobilise with growth factors alone in order to avoid toxicity, assuming the disease itself does not require chemotherapy at the time of attempted mobilisation.

Current Standard Agents

G-CSF (filgrastim, Amgen) has become the standard against which all other mobilisation agents are measured. This is because it has been shown to both mobilise more $CD34^+$ cells and have less toxicity than any other single agent against which it has been tested to date. It is not completely without toxicity, however, given that there have been deaths attributed to thrombosis (acute myocardial infarction and stroke) in sibling donors, possibly related to receptors on platelets for G-CSF.

Also, a recent study of serial ultrasounds in donors receiving G-CSF prompted by 4 reports of splenic rupture related to G-CSF demonstrated universal enlargement of the spleen during mobilisation, with regression of size after discontinuing the growth factor. Other effects of G-CSF which are shared with granulocyte-macrophage colony stimulating factor (GM-CSF) include pain, nausea, vomiting, diarrhea, insomnia, chills, fevers, and nightsweats.

GM-CSF (sargramostim, Berlex, Richmond, CA) as a single agent is used less often today for mobilisation than G-CSF, because it mobilises somewhat less well than G-CSF and because of a relatively higher incidence of both mild and severe side effects.

However, the fact that it can be more cost effective, and that there are reports of improved immune reconstitution with GM-CSF relative to G-CSF,, has caused some clinicians to reconsider its use. Further, for the patient, or normal donor who has failed to mobilise adequately on G-CSF alone, the combination of GM-CSF with or followed by G-CSF has been shown to be efficacious.,

Erythropoietin, now commonly used among cancer patients undergoing chemotherapy to maintain hemoglobin in the near normal range, also has some ability to mobilise $CD34^+$ cells. In the hard to mobilise patient, its use may therefore be doubly beneficial.

Alternative Agents

Stem cell factor (SCF) has been shown to be an excellent mobilising agent, particularly when used in combination with G-CSF. Unfortunately, the high incidence of allergic reactions to this agent and the need for observation after it is given have made it difficult to move into standard clinical practice. It is not currently available commercially in the US.

Newer Agents

Longer lasting variants of G-CSF (pegfilgrastim, Amgen) and erythropoietin (darbopoietin, Amgen) are now available and are in clinical trials as mobilising agents. They have the benefit of very long half-lives and so add an important measure of patient convenience and the probability that timing of collection may be more flexible without sacrificing optimal collections.

A new factor (AMD3100, AnorMed, Vancouver, Canada), which is a reversible inhibitor of the binding of stromal derived factor (SDF-1a) to its cognate receptor CXCR4, is currently in clinical trials as a mobilising agent. It is the first agent to be tried for mobilisation based on a rational understanding of its mechanism of action relative to HPC-stromal cell interactions. While it mobilises $CD34^+$ cells adequately on its own, it significantly improves the mobilisation capacity of G-CSF when used in combination with G-CSF in mice. Clinical trials in humans with various diseases are in progress, including myeloma.

Remobilisation

For the patient who fails to mobilise the necessary number of cells for transplant on the first attempt, but for whom it is clear that transplant is the best option, two decisions need to be made simultaneously: when to remobilise, and with what? When to remobilise is still a subject of debate, and depends to some extent on whether the failed mobilisation was with chemotherapy plus growth factor or growth factor alone.

Although at least one set of authors recommend immediate remobilisation with growth factors for a patient who has not mobilised adequately after chemotherapy plus growth factor, others feel that at least 2–3 weeks off growth factor prior to remobilisation offers the best chance of success., If adequate cells are given for an autologous transplant, it is even possible to collect enough for a second transplant during the period of white blood cell recovery following the first transplant. It is also possible to collect HPCs at least one year after a prior transplant: of 38 myeloma patients who attempted such a collection, 36 achieved an adequate number of cells for transplant (Cottler-Fox et al, in preparation).

What to use for a repeat mobilisation attempt for a specific patient and disease may be a complex decision. However, some general guidelines may be found in the literature. First, mobilisation with chemotherapy plus growth factor will generally yield more $CD34^+$ cells than growth factor alone. Second, chemotherapy plus G-CSF and SCF is more effective than chemotherapy plus G-CSF alone. Further, chemotherapy with either sequential or concurrent GM-CSF plus G-CSF may be more effective than G-CSF alone, as erythropoietin may improve the response to G-CSF.

Finally, retrospective analysis of data in myeloma patients from a single institution has shown that for patients who have undergone as many as 4 attempts at mobilisation, only 1 of the 4 may yield an adequate collection (Cottler-Fox et al, in preparation). It is possible that AMD3100 will find a niche in this hard to mobilise population: in an ongoing trial at the University of Arkansas for Medical Sciences in myeloma patients who have previously failed to collect at least 5×10^6 $CD34^+$ cells/kg, it provides adequate mobilisation in a significant number of patients (Tricot et al, in preparation).

An Algorithm for Mobilising Myeloma Patients

Although many variables may affect the decision of how and when to mobilise a myeloma patient, the following is a generally useful approach. Attempt first collection relatively early, i.e., with < 12 months prior therapy.

Attempt collection first with chemotherapy plus growth factor (except for patients over 70 with > 12 months prior therapy and platelets < 200 × 10^9/L, for whom growth factors alone may be tried first). The choice of growth factor (G-CSF versus GM-CSF) may depend on data under development regarding the importance of early immune reconstitution on time to progression and long-term disease-free survival. Pegylated filgrastim may replace standard G-CSF if studies show it to be equivalent or better than standard G-CSF.

If mobilisation is inadequate and patient is in CR or near-CR: wait at least 3 weeks, then remobilise with combination growth factors. In the future, AMD3100 may be a possibility.

If patient is not progressing, 3a and/or 3b may be repeated until adequate cells are collected.

If patient is progressing, and adequate cells are available for autologous transplant (e" 3 × 10^6 $CD34^+$ cells/kg), monitor $CD34^+$ cells in the blood at the time of leukocyte recovery, and consider collecting HPCs if there is adequate mobilisation.

If patient achieves a CR with transplant but does not mobilise adequately to collect during the leukocyte recovery phase post-transplant, consider attempting collection with combination growth factors at least 1 year after transplant.

General Principles

Although early attempts to use unmobilised peripheral blood stem cells (PBSCs) for autologous stem cell transplantation were problematic, the use of cytokines such as G-CSF to enhance the peripheralisation of $CD34^+$ cells and the collection of these stem cells using leukapheresis procedures has become the standard for autologous stem cell transplantation around the world. IBMTR and EBMT data suggest that over 80–90% of all autologous stem cell transplants in the world are performed using cytokine or chemotherapy/cytokine mobilised PBSCs as a source of stem cells.

In addition to reducing patient morbidity, the use of mobilised PBSCs has resulted in higher CD34 content of grafts, shorter hospital stays, and reduced engraftment times for both neutrophils and platelets as well as improved lymphocyte recovery resulting in enhanced immunologic reconstitution when compared to patients receiving autologous BM., These beneficial effects of mobilised PBSCs as a source of stem cells for autologous stem cell transplantation have been confirmed in a number of randomised trials.,

Based on the sustained success of using mobilised PBSCs for autologous stem cell transplantation, investigators began to pilot the use of PBSCs for allogeneic stem cell transplantation. Initial concerns focused on the possibility of increased risk of acute and chronic GVHD due to the presence of 10- to 50-fold increased numbers of T cells present in mobilised PBSC products.

It was not clear if the function of mobilised allogeneic T cells might be qualitatively altered resulting in even greater risk of GVHD or relapse. In addition, the risk of infusing increased numbers of cytomegalovirus (CMV)-positive granulocytes, dendritic cells, and monocytes into both CMV^- and CMV^+ recipients remained unknown and potentially posed an increased risk to the recipient. On the other hand, mobilised PBSCs contain 3 to 4 fold more $CD34^+$ cells, which might result in faster engraftment and more efficient transformation to complete donor chimersim.

Early Phase 2 studies demonstrated that G-CSF had a generalised effect on the peripheralisation of many different types of allogeneic peripheral blood cells, not just $CD34^+$ cells. These data are consistent with the notion that G-CSF has a generalised effect on remodelling the BM microenvironment, which results in the egress of many types of cells including T cells and monocytes. Activation of neutrophils by cytokines such as G-CSF results in the release of proteases that facilitate the egress of HSCs from the BM microenvironment. Interruption of the G-CSF signal through genetically "knocking out" the G-CSF receptor results in not only the expected elimination of G-CSF-induced mobilisation of HSCs but also IL-8 and chemotherapy-induced HSC mobilisation in these G-CSF receptor knock-out mice. Although it is not completely clear which is the most important tether binding HSCs to the microenvironment, LFA-1, VLA-4, CXCR4, and c-kit have all been implicated as critical stem cell adhesion molecules.,

Likewise, a number of neutrophil-specific enzymes have been implicated in mediating critical cleavages that result in stem cell egress from the microenvironment. These include neutrophil elastase, cathepsin G, proteinase 3, gelatinase B (MMP-9), and other metalloproteinases., Recent evidence has strongly implicated CD26, a CD34-associated protease, as the prime protease that may cleave SDF-1 off the marrow microenvironment resulting in the release of $CD34^+$ HSCs into the periphery. Of interest, a collaborative effort of the Link, Simmons and Levesque laboratories have shown that mice deficient in MMPq, neturophil elastase and cathepsin-G and mice deficient in dipeptidyl peptidase I (CD26) all mobilised hematopietic precursors in response to G-CSF normally. These data question the role of neutrophil specific proteins in stem cell egress (Daniel Link, personal communication).

Factors Determining Mobilisation and Outcomes

In the mid- and late-1990s, a large number of small Phase 2 studies were performed using mobilised PBSCs as a source of HSCs for allogeneic stem cell transplantation. All of these studies yielded similar results. Although neutrophil and platelet recovery was enhanced using cytokine mobilised allogeneic PBSCs, rates of acute GVHD were similar or less than that documented for BM as a source of allogeneic HSCs. The majority, but not all of these early studies, demonstrated increased actuarial rates of limited and extensive, chronic GVHD (cGVHD). Cost and hospitalisation appeared to be reduced compared to patients transplanted using allogeneic BM in these small Phase 2 studies.

Similar to many autologous PBSC studies, the number of allogeneic CD34 cells infused correlated well with both neutrophil and platelet engraftment. Brown demonstrated that those allogeneic PBSC recipients who had $> 5 \times 10^6$ CD34/kg infused had a 95% chance of both neutrophil and platelet engraftment by day +15. In this study, no correlation could be found between GVHD or survival and the number of $CD3^+$ cells infused. In one retrospective study by the MD Anderson group, infusion of $> 8 \times 10^6$/kg CD34 resulted in decreased survival presumably due to increased rates of cGVHD associated complications. These data have not been corroborated by other groups. Therefore, the infusion of high numbers ($> 8 \times 10^6$ CD34/kg) of allogeneic stem cells remains a controversial negative predictor for outcomes after allogeneic PBSC transplantation.

Effects on Donors

A major question is whether allogeneic stem cell harvesting results in less morbidity than BM harvesting for allogeneic stem cell donors. Anderlini et al reviewed 1448 mobilised allogeneic PBSC collections from the IBMTR and EBMT registries. G-CSF was used in > 99% of donors. Similar to multiple Phase 2 studies, 20% of normal donors required placement of a central line for apheresis. Eleven percent required more than 2 leukapheresis procedures and "serious complications" occurred in 1.1% of allogeneic PBSC collection versus 0.5% after BM harvest. Rowley et al[31] utilised an 11 point scale (0 = minimum and 10 = maximum) for 23 different symptoms occurring during the first 14 days after either BM or PBSC collection. There were no statistical differences between the PBSC and BM groups for any of the symptom complexes or for the severity of any symptoms.

Randomised Studies

Many of the Phase 2 studies assessing the role of mobilised allogeneic PBSCs on GVHD, relapse and overall survival are limited by design (Phase 2), paucity of patients, short-term follow-up, and heterogeneity of diseases for which allogeneic PBSC transplantation was performed. Several trials have provided insight into the relative effect of PBSC versus BM on GVHD, relapse, and survival. These include randomised studies, case control retrospective studies, and meta-analyses.

Summarises the results of all randomised trials comparing BM and PB as a source of stem cells for allogeneic stem cell transplantation., Half of these trials suffer from low numbers of patients. The 3 largest studies, demonstrate no significant difference in overall survival when peripheral blood is compared to BM as a source of allogeneic stem cells. These studies, similar to the smaller randomised trials, did demonstrate a significant enhancement in both neutrophil and platelet recovery consistent with the significantly increased numbers of CD34 cells that are harvested in mobilised allogeneic PBSC products compared with BM.

Randomised Trials Comparing Allogeneic Peripheral Blood to Bone Marrow

In one of the largest and best-designed studies, Bensinger et al, using identical conditioning regimens, GVHD prophylaxis (cyclosporine and methotrexate), and post-transplant growth factor support (no G-CSF), found a slight advantage in both disease-free and overall survival in those patients receiving mobilised allogeneic stem cells versus BM (P = .03 and P = .06, respectively). Although there was no difference in probability of 2 year overall survival in the subgroup of patients with less advanced disease (75% for PB and 72% for BM), those patients with more advanced disease demonstrated a significantly enhanced overall survival when mobilised allogeneic PBSCs were used as a source of stem cells (57% for PB and 33% for BM; P = .04). With a median follow-up for all surviving patients of 26 months (9–

47 months), the cumulative incidence of grade III–IV acute GVHD at 100 days was 64% in the PB group and 57% in the BM group (P = .35). The cumulative incidence of grade II–IV acute GVHD was 15% in the PB group and 12% in the BM group (P = NS). Although the follow-up was relatively short, the cumulative incidence of extensive cGVHD was 46% in the PB group and 35% in the BM group (P = .54). These results were inconsistent with many of the other smaller Phase 2 and Phase 3 studies, which all showed no difference in rates of acute GVHD and increased rates of cGVHD in recipients of allogeneic PBSC.

A recent meta-analysis was performed by Cutler et al summarising 15 Phase 2 and Phase 3 trials assessing the risk of GVHD in recipients of allogeneic PBSC and BM. This analysis demonstrated a modest increased relative risk of acute GVHD (relative risk 1.2) and a significant increased risk of developing cGVHD (relative risk 1.8) in recipients of allogeneic PBSCs. It also demonstrated a modest reduction in relative risk of relapse in recipients of allogeneic PBSCs compared to BM (relative risk 0.8).

Champlin et al and the IBMTR performed a retrospective case controlled study comparing the outcomes of recipients of allogeneic PBSCs and BM. Median follow-up was 1 year, and this study focused on 1 year outcomes. A total of 288 HLA-identical sibling PBSC recipients was compared with 536 case control allogeneic BM recipients. All patients received T-replete stem cell products. There was no significant difference in the incidence of grades II–IV acute GVHD (40% for PB and 35% for BM; P = NS) or grades III–IV acute GVHD (13% for PB and 19% for BM; P = NS). There was less variability in recovery times of both platelets and neutrophils after allogeneic PBSC compared to BM and statistically faster neutrophil and platelet recovery after allogeneic PBSC compared to allogeneic BM.

Treatment-related mortality (TRM), disease-free survival, and overall survival were similar for patients with acute leukemia in first remission. In contrast, acute leukemia patients in second remission and patients with CML in accelerated phase experienced lower TRM, improved DFS and overall survival when allogeneic PBSCs were used as a source of stem cells. There was no apparent difference in the risk of relapse after allogeneic PBSC versus BM transplantation. There was a trend toward lower relapse rates in patients with high-risk leukemia (acute leukemia in second remission and CML in accelerated phase). The relative risk of both limited and extensive cGVHD was increased in recipients of allogeneic PBSCs.

Mohty et al have performed the only long-term follow-up of allogeneic PBSC and BM recipients focusing specifically on the rates of cGVHD. At a median follow-up of 45 months (range 31–57 months), the 3 year cumulative incidence of cGVHD was 65% in the PBSC group (n = 53) and 36% in the BM group (n = 48) (P = .004). Extensive chronic GVHD was also more frequent in the PBSC group (44% versus 17%; P = .004).

These data suggest that although hematopoietic recovery is increased in recipients of allogeneic PBSC, there appears to be no increased risk of acute GVHD and a modest increase risk of cGVHD including extensive cGVHD. Overall survival in recipients of allogeneic PBSCs may be improved modestly but only in those patients with more advanced hematologic malignancies. Although no study has correlated rates of either acute or chronic GVHD with the number of CD3 cells/kg in the stem cell products, several studies have suggested that CD34 cells in excess of 8×10^6/kg found in allogeneic PBSC products are associated with a greater risk of both acute GVHD and GVHD.

Impact of G-CSF Mobilisation on Graft Content and Immune Reconstitution

Very little data exist on the relative impact of allogeneic PBSCs versus BM on immunologic reconstitution after allogeneic stem cell transplantation. Storek et al analysed the incidence of documented and suspected infections after transplantation of mobilised allogeneic PBSCs and BM in the randomised trial carried out by Bensinger et al. The cumulative incidence of infections was higher in the allogeneic BM group (120 versus 90 at 1 year). Since rates of acute GVHD after allogeneic PBSC infusions are similar to BM in spite of these products having 10 to 50 fold increased CD3 cells/kg over allogeneic BM products, a number of investigators have tried to understand the reason for this. To date, no studies have clearly shown a difference in either B-cell or T-cell recovery after allogeneic PBSC transplantation compared to BM transplantation.

A number of reports have emphasised the role of cytokines as mediators of GVHD. Cytokines produced by both CD4 and CD8 T cells can be segregated into two patterns: type I cytokines such as interferon-ã and IL-2 and type 2 cytokines, such as IL-4 and IL-10. Type 1 cytokines are proinflammatory and type 2 cytokines are considered anti-inflammatory. Multiple studies have shown that T cells that elaborate type 1 cytokines (Th1 cells) mediate GVHD whereas those

T cells that elaborate type 2 cytokines (Th2 cells) inhibit GVHD. Pan et al demonstrated that splenocytes from mice mobilised with G-CSF were polarised toward the Th2 phenotype. Those mice who received splenocytes from G-CSF-mobilised donor mice demonstrated significantly longer survival and less GVHD that those allogeneic transplant recipient mice who were infused with splenocytes from naïve unmobilised donor mice. T cells from G-CSF treated mice showed a significant increase in IL-4 production with a simultaneous decrease in IL-2 and interferon-ã production. This polarisation persisted in secondary mixed lymphocyte reactions (MLR) despite the absence of G-CSF during in vitro MLR.

Arpinati et al sized that G-CSF-mobilised PBSC contained antigen-presenting cells which prime T cells to produce Th2 cytokines. Two distinct lineages of dendritic cells (DC) have been described in humans. DC1 cells or myeloid DCs express HLA-DR, CD11c, CD13, and CD33 and require GM-CSF for their survival. These cells are negative for both myeloid and lymphoid specific markers ($Lin^{"}$), produce high levels of IL-12 when stimulated with tumour necrosis factor (TNF) or CD40 ligand and drive the differentiation of naïve T cells into the Th1 phenotype. DC2 or lymphoid DC are $HLA\text{-}DR^{+}/CD11c^{"}/CD4^{+}/IL\text{-}3Ra^{+}$ express high levels of T-cell receptor a chain and depend on IL-3, and not GM-CSF, for their survival and differentiation.

After appropriate activation, they can induce T-cell differentiation into Th2 cells. These investigators studied the effects of G-CSF mobilisation (10–16 ig/kg/day for 5 days) on DC content in the peripheral blood in these allogeneic donors. G-CSF treatment was found to mobilise DC2 but not DC1. Although the numbers of donors and controls studied were very small, the median number of DC1 per litre in the G-CSF group was not different (11 versus 10 × 10^{6}/litre; P = .52) than in the control premobilisation group. In contrast, the numbers of DC2 were significantly increased in the G-CSF mobilisation group compared with control (median 24.8 versus 4.9 × 10^{6}/litre; P = .0009).

As expected, a proliferative response of naïve allogeneic T cells could be detected in vitro to fresh DC1 but not to fresh DC2. Activation of DC2 in vitro with TNF, GM-CSF, and IL-3 resulted in the rapid maturation of these precursors to mature DC2 cells overexpressing costimulatory molecules such as CD34 and CD86 restoring their ability to induce a proliferative response to naïve $CD4^{+}$/CD45RA allogeneic T cells. Incubation of naïve allogeneic T cells with DC1 resulted in polarisation of these T cells toward the Th1 phenotype as measured by restimulation of these T cells with PMA and ionomycin and detecting primarily IL-2 and interferon-ã as the major intracellular cytokines produced after restimulation. In contrast, incubation of naïve T cells with DC2 cells results in the polarisation of these T cells toward the Th2 phenotype as noted by the intracellular accumulation of IL-4 and IL-10 after restimulation in vitro with PMA and ionomycin. In conclusion, G-CSF mobilisation results in stem cell products with 10- to 50-fold more T cells and 4- to 6-fold more DC2 cells.

The increased numbers of DC2 in G-mobilised products may reduce the relative risk of acute GVHD as seen in preclinical murine allogeneic transplant studies described above and observed in the initial Phase 2 and 3 clinical trials in humans comparing BM versus mobilised PBSCs as sources of allogeneic stem cells. It is of interest that cord blood stem cell products, which are associated with a low risk of severe acute GVHD, contain primarily DC2 and negligible DC1. Consistent with this notion, Waller found a strong inverse correlation with the number of precursor DC2 infused in allogeneic bone marrow and the incidence of both cGVHD and, more importantly, relapse.

Alternative Allogeneic PBSC Mobilisation Regimens

Although other cytokines, in addition to G-CSF, have been used to mobilised autologous PBSCs from humans including GM-CSF, Flt-3 ligand, stem cell factor (SCF), Daniplestim (IL-3 agonist), thrombopoietin agonists, chimeric cytokines including Leridistim (IL-3 agonist-G-CSF chimeric molecule) and Progenipoietin-1 (Flt-3 ligand-G-CSF chimeric molecule), peg-filgrastim (Neulasta™) and SDF-1 antagonist (AMD 3100), only G-CSF and GM-CSF have been approved by the FDA and only G-CSF has been studied extensively for the mobilisation of allogeneic PBSCs in humans.

We have performed several sequential nonrandomised trials to determine the comparative effects of allogeneic PBSCs mobilised with G-CSF at 10 ig/kg/day (n = 96), G-CSF (10 ig/kg/day) combined with GM-CSF at 5 ig/kg/day (n = 102) or GM-CSF alone (10 or 15 ig/kg/day (n = 32). A comparison of the various allogeneic PBSC mobilisation regimens is shown in All donors underwent leukapheresis (20 litre) on the fifth day of cytokine administration. The target CD34+ content was 5.0×10^6 CD34/kg with a minimum of 2×10^6/kg. The data shown in are expressed as mean +/" SD. PBSCs mobilised with G/GM resulted in collection of grafts with similar CD34 content compared with G alone. Grafts obtained following mobilisation with GM alone contained significantly fewer $CD34^+$ cells, but sufficient numbers for rapid engraftment. Grafts mobilised with GM or G/GM contained significantly fewer T and NK cells. There were no obvious differences in donor toxicities including bone pain. All recipients of GM mobilised cells engrafted with kinetics similar to recipients of G and G/GM, although neutrophil recovery was delayed about 1 day. Rates of neutrophil and platelet recovery and cGVHD for the G and GM groups are shown in In 30 evaluable recipients of peripheral blood grafts mobilised with GM alone, the actuarial risk of grades 2–4 acute GVHD was only 0.13 ± 0.05 and 0.00 (0/31) for grades 3–4 acute GVHD.

In a multivariate analysis including patient and donor age, sex mismatching, conditioning regimen received, $CD3^+$cell dose and $CD34^+$ cell dose, only the receipt of PBSCs mobilised with GM-CSF alone correlated with a lower risk of grades 2–4 acute GVHD. These data suggest that altering the mobilisation regimen and cytokines used may alter the functional aspects of the graft thereby modifying outcomes such as GVHD. Randomised Phase 3 studies will need to be performed to more accurately dissect the phenotypic differences in grafts mobilised by different cytokines/chemokines or combinations and to more accurately assess the impact on important endpoints such as multilineage engraftment, disease-free survival, overall survival, GVHD, and relapse.

Effect of Cytokine Mobilisation Regimen on Graft-versus-host Disease (GVHD)/Survival

Chemokines such as IL-8 have been used to induce the egress of hematopoietic stem cells into the peripheral blood of mice and nonhuman primates. This effect is rapid (30 minutes-4 hours) and may result from the ability of these chemokines to induce the release of a protease from mature myeloid cells resulting in a decrease in the intramedullary concentration of SDF-1, the ligand for the receptor CXCR4, which is expressed in many cells including hematopoietic stem cells.

Increasing evidence points to the critical role of the CXCR4/SDF-1 axis in both murine and human stem cell mobilisation. The bicyclam molecular AMD 3100 was first clinically developed for its potent and selective inhibition of HIV type 1 and 2 replication through binding to the chemokine receptor, CXCR4. Initial clinical trials in AIDS patients demonstrated that AMD-3100 induced a rapid (within 1 hour) increase in both WBC and circulating progenitor cells. Broxmeyer and colleagues demonstrated a 40-fold increase in the mobilisation of hematopoietic progenitors within 1 hour of AMD-3100 injection in mice. Studies in human volunteers and the first Phase 1 and 2 clinical trials in patients undergoing autologous stem cell transplantation are currently underway. These studies have shown a consistent and rapid impact on stem cell mobilisation when given alone and a synergistic effect on $CD34^+$ mobilisation when coadministered with G-CSF. Since CXCR4 is expressed on different types of cells including $CD34^+$ progenitor and T cells, the impact of AMD-3100 on both multilineage engraftment and GVHD will need to be appropriately explored in preclinical animal models before it can be safely used as a rapid mo

Synthetic biology:

The safety of early applications of synthetic biology may be adequately addressed by the existing regulatory framework for biotechnology, especially in contained laboratories and manufacturing facilities. But further advances in this emerging field are likely to create significant challenges for U.S. government oversight, according to a new report authored by Michael Rodemeyer of the University of Virginia. Synthetic biology promises major advances in areas such as biofuels, speciality chemicals, and agriculture and drug products. In New Life, Old Bottles: Regulating First-Generation Products of Synthetic Biology, Rodemeyer examines the benefits and drawbacks of using the existing U.S. regulatory framework for biotechnology to cover the new products and processes enabled by synthetic biology. According to Rodemeyer, initial synthetic biology products will be relatively simple modifications of current technology and can be addressed by existing biotechnology regulations with only modest revisions.

However, as the technology develops, regulatory agencies such as the Environmental Protection Agency and Food and Drug Administration will face challenges in assessing potential risks and the adequacy of controls, especially if complex synthetic microorganisms are released into the environment. Today's risk assessment practices and laws like the Toxic Substances Control Act and Federal Food, Drug, and Cosmetic Act, simply are not designed to handle 21st century technological advances.

"Before synthetic biology matures, Congress and policymakers should consider how to rationalise and modernise the regulation of new converging technologies, instead of attempting to shoehorn each new area of technological development into laws previously written for a different set of issues and potential risks," Rodemeyer argues.

"It would be easy to relegate discussions about oversight to the backburner. But procrastination bears a risk. A productive dialogue may become more difficult as synthetic biology evolves and stakeholders become divided in their opinions about benefits and risks. The existing regulatory framework for biotechnology is the natural starting point for synthetic biology oversight. But the framework is at best a patchwork quilt of decades old guidelines and laws that could impede innovation, undercut public confidence, and compromise the promised benefits of synbio," says David Rejeski, the director of the Foresight & Governance Project at the Woodrow Wilson International Centre for Scholars. "Policymakers, industry, and other key stakeholders should start a discussion now on the basic question of whether existing regulations will work with advanced synthetic biology, and if not, what changes may be needed to ensure safe devel Synthetic Biology: funders move to address social and ethical challenges

The report of an independent review of social and ethical challenges associated with research into, and the application of, synthetic biology, is published today (9 June). The report, commissioned by the Biotechnology and Biological Sciences Research Council's (BBSRC) Bioscience for Society Strategy Panel, is part of its programme to ensure that BBSRC adequately addresses issues raised by this rapidly emerging area of science and technology. Embargoed copies of the report are available upon request. Synthetic biology seeks to apply the principles of engineering to biological systems and processes. Scientists believe that it may lead to new applications, such as new energy production systems, medical therapies, biological computers and innovative ways to clean up hazardous waste. In common with other modern technologies, it is potentially controversial because it raises issues of ownership, misuse, unintended consequences, and accidental release.

The report, 'Synthetic Biology: social and ethical challenges', has been written by Andrew Balmer and Paul Martin of the Institute for Science and Society, University of Nottingham, a leading research centre working on the impact of new technology. It reviews what synthetic biology is, where it has come from, and where it is going, as well as making recommendations to research funders and the scientific community about how social and ethical issues should be addressed. These include:

- The need for scientists to engage with the public early in the development of synthetic biology to ensure that research does not get ahead of public attitudes.
- Synthetic biology must not be over-hyped by its supporters and critics should not exaggerate the risks it poses.
- Reviewing current regulations and guidelines to ensure that an appropriate governance framework is in place before most synthetic biology applications are introduced.

BBSRC, with advice from its Bioscience for Society Strategy Panel, is considering the recommendations of the report and will use its conclusions to inform its future policy in this area.

"Synthetic biology is in the early stages of development in the UK, and it is an appropriate time to address issues of public interest and concern," said Professor Nigel Brown, BBSRC Director of Science and Technology, "BBSRC already requires its grantholders to address ethical and other social issues, but this report will help us to focus on those concerns associated with synthetic biology."

BBSRC is already working with its sister Research Councils on ethical and regulatory issues that might arise from synthetic biology research. BBSRC is working closely with the Engineering and Physical Sciences Research Council and the Royal Society on how to take forward public dialogue and engagement on the science of synthetic biology, and with the Economic and Social Research Council and the Arts and Humanities Research Coun Scientific Grand Challenges identified to address global sustainability.

The international scientific community has identified five Grand Challenges that, if addressed in the next decade, will deliver knowledge to enable sustainable development, poverty eradication, and environmental protection in the face of global change. The Grand Challenges for Earth system science, published today, are the result of broad consultation as part of a visioning process spearheaded by the International Council for Science (ICSU) in cooperation with the International Social Science Council (ISSC). The consultation highlighted the need for research that integrates our understanding of the functioning of the Earth system—and its critical thresholds—with global environmental change and socio-economic development.

The five Grand Challenges are: 1. Forecasting—Improve the usefulness of forecasts of future environmental conditions and their consequences for people. 2. Observing—Develop, enhance and integrate the observation systems needed to manage global and regional environmental change. 3. Confining—Determine how to anticipate, recognise, avoid and manage disruptive global environmental change. 4. Responding—Determine what institutional, economic and behavioural changes can enable effective steps toward global sustainability. 5. Innovating—Encourage innovation (coupled with sound mechanisms for evaluation) in developing technological, policy and social responses to achieve global sustainability.

'The challenges are a consensus list of the highest priorities for Earth system research and provide an overarching research framework. If we, the scientific community, successfully address these in the next decade, we will remove critical barriers impeding progress toward sustainable development,' said Dr Walt Reid, who chaired the Task Team overseeing the first step of the visioning process.

'Addressing these challenges will require new research capacity, especially the involvement of young scientists and scientists from developing countries, and a balanced mix of disciplinary and interdisciplinary research that actively involves stakeholders and decision makers,' Dr Reid continued.

'The existing global environmental change programmes Diversitas, International Geosphere Biosphere Programme, International Human Dimensions Programme and the World Climate Research Programme along with the Earth System Science Partnership have played an important role in our understanding of the Earth system,' explained Professor Johan Rockström, the current chair of the visioning Task Team.

'Their engagement has been an important part of the visioning process and the continued involvement of these global research networks is essential to the globally coordinated research effort needed to address the questions posed by the Grand Challenges,' Professor Rockström said.

Now that the research framework has been identified the next step has begun: determining the organisational structure required to implement this framework.

Professor Deliang Chen, ICSU Executive Director, said: 'A lot of integrated research is already happening but it does not constitute the concerted coordinated global effort that is needed to effectively respond to the Grand Challenges. ICSU, together with the ISSC and the Belmont Forum of funders, are consulting with the existing programmes and related initiatives to determine what new structure(s) will be required.'

'The new structure(s) will need to deliver the science to answer the Grand Challenges more rapidly and more effectively than is likely to happen with the current arrangements,' Professor Chen concluded.

Finnish Researchers Discover Regulator of Human Cell Aactivity

Research teams headed by Prof. Johanna Ivaska (University of Turku and VTT Technical Research Centre of Finland) and Prof. Marko Salmi (University of Turku and the National Institute for Health and Welfare) have discovered that the SHARPIN protein regulates human cell activity. Published in *Nature Cell Biology*, the study concludes that SHARPIN regulates the movement and activity of inflammatory cells and of lung and prostate cancer cells. It is likely that the discovery will also have significant implications for other conditions such as Crohn's disease, psoriasis, rheumatism and even MS.

The hypothesis that there must be a protein that inhibits cell activity was posited as far back as in the early 1990s. Researchers have now shown that SHARPIN is the sized protein.

Juha Rantala and Jeroen Pouwels, members of Prof. Ivaska's research team, discovered the role of SHARPIN using a cell chip screening method developed at VTT. Using this metho Dramatic diversity of columbine flowers explained by a simple change in cell shape.

Columbine flowers are recognisable by the long, trailing nectar spurs that extend from the bases of their petals, tempting the taste buds of their insect pollinators. New research at Harvard and the University of California, Santa Barbara (UCSB) helps to explain how columbines have achieved a rapid radiation of approximately 70 species, with flowers apparently tailored to the length of their pollinators' tongues.

Bees, for example, enjoy the short spurs of *Aquilegia vulgaris*, whereas hawkmoths favour *A. longissima*, whose spurs can grow to up to 16 centimetres.

According to results published November 15 in the *Proceedings of the Royal Society B*, the dramatic diversity in the length of the columbines' spurs is the result of one simple change during development: the extent of cell elongation.

"The evolutionary importance of interactions between flowers and pollinators has been recognised for centuries," says co-lead author Sharon Gerbode, a postdoctoral fellow at the Harvard School of Engineering and Applied Sciences (SEAS).

"Charles Darwin, observing orchids, recognised that the extraordinarily long nectar spur on the *Angraecum* must have evolved in concert with the equally long tongue of the moth that pollinated it, but the exact mechanism for this kind of adaptation has been a matter of speculation."

For more than 60 years, biologists have assumed that the length of columbine spurs was achieved primarily by cell proliferation. The new research reveals that proliferation plays almost no role at all in creating the vast diversity of *Aquilegia* species currently seen.

In fact, 99 percent of the variation in spur length can be attributed to changes in cell shape specifically, changing round cells into long ones.

The researchers made more than 13,000 measurements to count the number of cells along the spur, as well as the area and degree of elongation of each cell.

They found that cell division ceases early in the development of the spur when it is about 5 millimetres long. At that point, the general pattern for the spur has been established, and all species of columbine petals look the same. From that point on, the cells elongate to varying extents, creating diverse spur lengths across species.

"The controlled elongation of cells within the petal spurs was a critical evolutionary innovation for *Aquilegia*, a genus that is considered to be a textbook example of adaptive radiation," says co-lead author Joshua Puzey, a graduate student in Harvard's Department of Organismic and Evolutionary Biology (OEB).

The researchers confirmed their results through mathematical analysis and modelling, and through *in vivo* experiments to disrupt cellular structure. The next step will be to examine several major hormone pathways and cytoskeletal elements that are known to influence cell elongation and developmental timing.

"We want to understand the genes that are responsible for both the initial evolution of nectar spurs and their subsequent diversification," says co-principal investigator Elena Kramer, Professor of OEB at Harvard.

It is clear, she says, that the starting point for the spur is likely to have already been present in the last common ancestor of all the columbine species.

"Now that we understand the real developmental basis for the first appearance and diversification of spurs, we can make much more informed guesses about what genes contributed to the process," Kramer adds.

"Fundamentally, these studies will help us answer questions about the genetic basis for speciation and how developmental processes evolve."

Columbines show promise as a model organism for the study of evolution in plants because they have experienced such a rapid adaptive radiation within the past 3 million years.

"The fact that this occurred quite recently is incredibly useful," says Kramer, "because it means that the species are still very similar to each other at the genetic level."

Once researchers have identified the molecular signals that drive elongation in the spurs, the hope is that they will be able to recognise and understand speciation at all levels, from genes to populations.

"*Aquilegia* serve as a nice example of how environmental selective pressures may drive extreme morphologies as here the flower and pollinator strive for an exclusive relationship," adds co-principal investigator L. Mahadevan, the Lola England de Valpine Professor of Applied Mathematics at SEAS and Professor of OEB and Physics at Harvard.

"Given that we can now manipulate spur length using externally applied drugs, our study even raises the possibility of artificially tuning that process and studying the results from an ecological perspective."

The research was supported by the MacArthur Foundation, the Wyss Institute for Biologically Inspired Engineering at Harvard, The Kavli Institute for Bionano Science and Technology at Harvard, the National Science Foundation (NSF), and the NSF-supported Materials Research Science and Engineering Centre at Harvard.

Michigan State University Study Sheds Light on Microscopic Dlower Petal Ridges

Microscopic ridges contouring the surface of flower petals might play a role in flashing that come-hither look pollinating insects can't resist. Michigan State University scientists and colleagues now have figured out how those form. The result could help researchers learn to enhance plants' pollination success and even could lead to high-grip nanomaterials and "green chemical" feedstocks.

"Surprisingly, our work on plant surface biochemistry became a birds and bees and flowers story," said John Ohlrogge, MSU University Distinguished Professor of plant biology. "It's a fundamental property of plant flowers, and we've discovered a basis of how these ridges are made."

Known for 75 years, the exact biological function and nature of the flower nanoridges still eludes scientists. They might help pollinating insects grip petals, and retain glistening water droplets that could attract the visitors. Because the ridges' spacing is approximately that of visible and ultraviolet light wavelengths, moreover, some recent research suggests they produce an iridescent shimmer that attracts pollinators.

To start, visiting professor Mike Pollard and former Ohlrogge post-doctoral research associates Fred Beisson and Yonghua Li tapped new genetic information to find a mutated strain of the standard research plant Arabidopsis thaliana mustard weed. The petals have no such nanoridges because the mutation inhibits production of a polymer that forms the plant cuticle, which separates cell walls from plants' waxy surfaces.

Examining the mutant plants' flowers and comparing them to normal mustard plants under scanning electron microscopes, the researchers found that the ridges form from cutin polyester, not from underlying surfaces as some have speculated. How that occurs from surface folding or uneven synthesis of cutin polymer across the cell wall, for example, has yet to be learned.

But the research will open doors to further research based on cuticular nanostructures, the researchers noted in latest recent edition of the journal *Proceedings of the National Academy of Sciences.*

"That could include production of polyesters or related basic chemicals by genetically manipulating plants or microbes" said Beisson, now at Aix Marseille Université in Saint-Paul-lez-Durance, France.

CSHL Neuroscientist Discovers Protein that Regulates Forgetting of Short-term Memories

Memories that we have just acquired – a new phone number, or the name of a new acquaintance are more liable to be forgotten than memories we have held for some time. We know this from experience, but we are just learning about events inside and between nerve cells that account for the loss of short-term memory. Now, a neuroscience team led by a scientist at Cold Spring Harbor Laboratory (CSHL) has discovered that three kinds of forgetting – all involving the erasure of short-term memory – are regulated within neurons by the activity of a protein called Rac. The results will be published online ahead of print February 19 in the journal *Cell*.

Naturally occurring Rac activation within neurons located in a fruit fly organ called the mushroom body was linked experimentally with three distinct memory-erasure tasks by CSHL Professor Yi Zhong, Ph.D., and colleagues at Tsinghua University in Beijing. One kind of erasure is associated with gradual short-term memory loss (previously considered "passive"); another entails an acute, rapid removal of short-term memory (e.g., "reversal learning"); a third involves a kind of erasure associated with new information that interferes with an existing short-term memory ("interference").

In all three kinds of erasure, the team has found, the process of forgetting is mediated by a mechanism dependent upon the activity level of Rac, a protein that belongs to the Rho family of GTPases. These are a type of protein known to act, among other things, as regulators of the cytoskeleton, the superstructure of cells. Importantly, Zhong and colleagues propose that Rac's role in erasing memory is directly related to its function as a cytoskeleton remodelling agent.

In genetically modified flies, the team elevated Rac activity in mushroom body neurons, which are the seat of olfactory-based memories in the fly. Elevated Rac activity had the effect of accelerating the decay of short-term memories. The process worked in the other direction, too. By inhibiting the activity of Rac they observed that short-term memories decayed more slowly than normal, lasting more than a full day rather than vanishing within three hours, as they do, on average, when Rac is not blocked.

The scientists reported that the Rac-dependent forgetting mechanism in flies did not affect mechanisms involved in the formation of new memories, specifically a pathway associated with the gene Rutabaga that is mediated by the enzyme adenylyl cyclase.

"The molecular basis of short-term memory really has been overlooked by the neuroscience community," says Zhong, a CSHL professor. "It has been widely assumed that such memory is degraded through passive cellular processes. Our experiments challenge the notion by providing evidence of a dedicated mechanism for removing several kinds of short-term memories."

The active forgetting mechanism that Zhong and colleagues have uncovered has potentially adaptive benefits for an organism, they speculate. "The Rac-mediated mechanism makes possible a forgetting strategy that can respond to environmental information," Zhong explains. "For instance, it could be advantageous to forget a memory in disuse that is no longer of biological significance, or a memory inconsistent with current circumstances."

"Forgetting Is Regulated through Rac Activity in Drosophila" appears online ahead of print in *Cell* February 19. The authors are: Yichun Shuai, Binyan Lu, Ying Hu, Lianzhang Wang, Kan Sun and Yi Zhong.

CSHL Study Finds Short and Long-term Memories Require Same Gene but in Different Circuits

Why is it that you can instantly recall your own phone number but have to struggle with your mental Rolodex to remember a new number you heard a few moments ago? The two tasks "feel" different because they involve two different types of memory long-term and short-term, respectively that are stored very differently in the brain. The same appears to be true across the animal kingdom, even in insects such as the fruit fly. Assistant Professor Josh Dubnau, Ph.D., of Cold Spring Harbor Laboratory (CSHL) and his team have uncovered an important molecular and cellular basis of this difference using the fruit fly as a model. The results of their study appear in the August 25 issue of *Current Biology*.

The CSHL team has found that when fruit flies learn a task, each of two different groups of neurons that are part of the centre of learning and memory in the fly brain simultaneously forms its own unique memory signal or trace. Both types of trace, the team discovered, depend on the activity of a gene called rutabaga, of which humans also have a similar version. A rapidly occurring, short-lived trace in a group of neurons that make up a structure called the "gamma" (ã) lobe produces a short-term memory. A slower, long-lived trace in the "alpha-beta" (aâ) lobe fixes a long-term memory.

A Tale of Two Lobes

Neuroscientists call the rutabaga gene a coincidence detector because it codes for an enzyme whose activity levels get a big boost when a fly perceives two stimuli that it has to learn to associate with one another. This enzymatic activity in turn signals to other genes critical for learning and memory.

A classic experiment that teaches flies to associate stimuli – and one that the CSHL team used – is to place them in a training tube attached to an electric grid, and to administer shocks through the grid right after a certain odour is piped into the tube.

Flies with normal rutabaga genes learn to associate the odour with the shock and if given a choice, buzz away from the grid. But flies that carry a mutated version of rutabaga in their brains lack both short- and long-term memory, don't learn the association, and so fail to avoid the shocks.

The team has now found, however, that this total memory deficit does not occur when flies carry the mutated version in either the ã or in the aâ lobes. Flies in which normal rutabaga function was restored within the ã lobe alone regained short-term memory but not long-term memory. Restoring the gene's function in the aâ lobe alone restored long-term memory, but not short-term memory.

Long- and Short-term Memory Involve Different Circuits

"This ability to independently restore either short- or long-term memory depending on where rutabaga is expressed supports the idea that there are different anatomical and circuit requirements for different stages of memory," Dubnau explains. It also challenges a previously held notion that neurons that form short-term memory are also involved in storing long-term memory.

Previous biochemical studies have suggested that rapid, short-lived signals characteristic of short-term memory cause unstable changes in a neuron's connectivity that are then stabilised by slower, long-lasting signals that help establish long-term memory in the same neuron. But anatomy studies have long hinted at different circuits. Surgical lesions that destroy different parts of an animal's brain can separately disrupt the two kinds of memory, suggesting that the two memory types might involve different neuronal populations.

"We've now used genetics as a finer scalpel than surgery to reconcile these findings," Dubnau says. His team's results suggest that biochemical signalling for both types of memory are triggered at the same time, but in different neuron sets. Memory traces form more quickly in one set than the other, but the set that lags behind consolidates the memory and stores it long-term.

Why Two Mechanisms?

But why might the fly brain divide up the labour of storing different memory phases this way? Dubnau's hunch is that it might be because for every stimulus it receives, the brain creates its own representation of this information. And each time this stimulus – for example, an odour – is perceived again, the brain adds to the representation and modifies it. "Such modifications might eventually disrupt the brain's ability to accurately remember that information," Dubnau speculates. "It might be better to store long-term memories in a different place where there's no such flux."

The team's next mission is to determine how much cross talk, if any, is required between the two lobes for long-term memory to get consolidated. This work will add to the progress that scientists have already made in treating memory deficits in humans with drugs aimed at molecular members of the rutabaga-signalling pathway to enhance its do Extensive protein interaction network controls gene regulation

The genes of a cell are like the 88 keys of a piano. To play chords and music, however, the keys must be activated in exact combinations by a pianist's hands. Those hands represent the coregulators of a cell that simultaneously and precisely activate genes to produce all of the cell's functions. More than half of your DNA is devoted to regulating how the genes that make proteins – the workhorses of the cells – carry out their tasks, said Dr. Bert O'Malley who, with Dr. Jun Qin co-led a team of scientists at Baylor College of Medicine that over eight years identified and classified virtually all the transcriptional coregulators in a human cells. These coregulators – coactivators and corepressors – control how and to what degree genes are turned on or off as well as when they are active and for how long. The more than 11,000 coregulators identified – the focus of O'Malley's work for more than 15 years – form and act in approximately 3,000 multi-protein complexes that function in the human cell. A report on their work appears in the current issue of the journal *Cell.*

"Genes are how we inherit our capacities," said O'Malley, chair of molecular and cellular biology at BCM and a National Medal of Science recipient. "The DNA functions by first coding for the synthesis of RNAs (another form of genetic material), which in turn, directs the synthesis of proteins in the cells. Proteins are the final functional units emanating from the genes. They carry out all the biochemical reactions needed for a cell to live, grow and function. Coregulators are the helper proteins that actually decode the information in our genes."

"Surprisingly, we found that over half of our total DNA is used simply to create the immense number of coregulators that, in turn, regulate the expression of our genes. This indicates that 'precise regulation' in decoding genes is an absolutely mandatory rule in human cells, and that this occurs via the coregulator proteins," he said.

Dr. Ronald Margolis, senior advisor for molecular endocrinology at the National Institute of Diabetes and Digestive and Kidney Diseases the division within the National Institutes of Health that supported the research—said the findings provide an important new tool for further research into endocrine and metabolic diseases such as diabetes and osteoporosis.

"Ultimately this work gives us new and important insights that are key to understanding how and why all types of hormones work the way they do," he said. "It's just this kind of basic research that provides the foundation for new diagnostics, therapies and devices."

Qin, professor in the departments of biochemistry and molecular biology and molecular and cellular biology at BCM, is a world expert in mass spectrometry, the backbone technique that enabled the scientists to identify and analyse the proteins and protein complexes.

Qin said the vision of Drs. O'Malley and Adam Kuspa, chair of biochemistry and molecular biology at BCM, enabled him to take the bold step of analysing these proteins and determining how they work together – a massive project that provides a blueprint of knowledge on which to build new understanding of how proteins work and how their malfunctions result in disease.

"A curious journey sometimes can land on the right place," said Qin.

He credits Dr. Anna Malovannaya, who came to his laboratory as a graduate student, with developing the techniques that made it possible not only to identify the proteins but to figure out which ones work together and how.

"Determining the composition of the proteome the entire set of proteins produced by a genome does not tell you how it all works," said Malovannaya.

"Proteins work in groups. This study does not just profile them. It also tells how they interact with one another," Qin said.

"The way we looked at it was new," said Malovannaya. "We had to build new tools for grouping proteins into functional complexes and figure out which ones were important."

Achieving that took thousands of experiments. When they had done about 1,000 experiments, the answers became clearer, said Qin.

The Nuclear Receptor Signalling Atlas (NURSA) was the catalyst for the work, said Qin. O'Malley and Dr. Ronald Evans of the Salk Institute are co-directors of the project that is funded by the National Institute of Diabetes and Digestive and Kidney Diseases.

O'Malley and his colleagues were surprised by the fact that more than half of all human DNA goes into producing the coregulators that decode genes, but in retrospect, it makes sense. They expected to find about 500 genes for directing the synthesis of coregulators and, instead, identified more than 11,000.

"The regulation of gene expression is complex," O'Malley said. "It is critical that genes turn on at the right time, in the exact right amount and under the right condition. If a gene makes 10 percent too much or too little of a protein, then the person develops a disease or functions poorly."

"It's all about accurate regulation and combinatorial regulation," he said. "Many hundreds of genes must be regulated together at precisely the same time. The cell is a master at that. Every gene has to function perfectly for a cell to work correctly – and the coregulators make it happen. It is one of the most amazing events biologists have discovered – beautifully complex and fine-tuned."

The eight-year project is of a magnitude similar to that of sequencing the genome, but "now we have determined the composition of the coregulator proteome," said O'Malley. Synthesis of the proteome is directed by the genome. Which proteins are produced and at what time depend on the type of cell and the functions of its coregulators. Malovannaya and Dr. Rainer B. Lanz, assistant professor, both in the department of molecular and cellular biology, were first authors of the *Cell* paper and contributed equally to the research.

Stress can Control Our Genes

Stress has become one of the major disease states in the developed world. But what is stress? It depends on from where you look. You may experience stress as something that affects your entire body and mind, the causes of which are plentiful. But if we zoom in on the building bricks of the body, our cells, stress and its causes are defined somewhat differently. Stress can arise at the cellular level after exposure to pollution, tobacco smoke, bacterial toxins etc, where stressed cells have to react to survive and maintain their normal function. In worst case scenario, cellular stress can lead to development of disease. Researchers from Dr.Klaus Hansen's group at BRIC, University of Copenhagen, have just shown that external factors can stress our cells through the control of our genes. "We found that stress-activating factors can control our genes by turning on certain genes that were supposed to be silenced.

It is very important that some genes are on and others are off in order to ensure normal foetal development and correct function of our cells later in life" says Dr. Klaus Hansen. Simmi Gehani, PhD-student in the Hansen group, found that exposing human cells to a stress-activating compound turned on silenced genes. Even brief changes in gene activation can be disastrous during foetal development as establishment of correct cellular identity can be disturbed in our cells. But altered gene activity can also have consequences in the adult body. "For example, one could imagine that prolonged stress causes nerve cells in the brain to produce hormones and other signalling molecules they do not normally produce and this can disturb normal brain function" says Simmi Gehani.

The Hansen research group is very interested in understanding how our genes are turned on and off. "We know that different protein complexes can associate with specific proteins (histones) to which DNA is wound around and thereby determine whether the genes are active or inactive. Small chemical groups can cause protein complexes to bind to histones and these can control gene activity" says Dr. Klaus Hansen. The researchers have studied in detail a complex called PRC2. PRC2 can attach small chemical groups - methyl groups - to the histones. Protective complexes can bind to the histones when this marker is present and the genes are turned off.

Their new results show that the protective complexes are lost and selected genes turned on when cells are exposed to external stress factors. The reason why the complexes are lost is that the stress factors instruct an enzyme named MSK to attach another chemical group - a phosphate group - to the histones neighbouring the methyl group. The phosphate group neutralises the effect of the methyl group and turns specific genes on. "The consequence is that genes that should be turned off are now active and this may disturb cellular development, identity and growth" says Simmi Gehani. This means that without damaging our genetic code external stress factors can control the activity of our genes. The results are published today in the renowned international journal *Molecular Cell.*

New Regulator of Circadian Clock Identified

Montreal, October 20, 2010 – Daily sleeping and eating patterns are critical to human well-being and health. Now, a new study from Concordia University has demonstrated how the brain chemical dopamine regulates these cycles by altering the activity of the "clock-protein" PER2. Published in the *Journal of Neuroscience*, these findings may have implications for individuals with Parkinson's Disease with disrupted 24 hour rhythms of activity and sleep.

"PER2 is a protein well-known for its role in the regulation of daily or circadian rhythms, this is why it is referred to as a clock protein," says senior author, Shimon Amir, a psychology professor at the Concordia Centre for Studies in Behavioural Neurobiology. "Many molecules, such as stress hormones are known to have an impact on the activity of PER2. Until now, the role of dopamine in regulating circadian rhythms has been unclear. Our findings show that not only is PER2 influenced by dopamine but also that dopamine is necessary for its rhythmic expression in specific brain regions."

Dopamine and Parkinson's

Parkinson's disease is caused by the degeneration of specific nerve cells, which results in a decrease in dopamine levels in the brain. Dopamine is critical for normal movements and balance and its decreased level results in instability and involuntary movements, the telltale symptoms of Parkinson's.

The findings from this Concordia study may explain the disruptions of daily behavioural and physiological rhythms that are also frequently reported in Parkinson's.

Rise in Dopamine Followed by Rise in PER2

Amir and his colleagues studied the role of dopamine in rats. Their first steps were to show that PER2 is present in a specific brain area that normally receive dopamine, namely the dorsal striatum, and that it fluctuates daily in this area.

In this same region of the brain the research group demonstrated that a rise of dopamine preceded the rise in PER2 and that removing dopamine from the brain or blocking one of its receptors resulted in decreased PER2, which, in turn, could be reversed by the administration of a drug that mimics the action of dopamine on this receptor.

"Our findings are consistent with the idea that the rhythm of expression of PER2 depends on the daily activation by dopamine," says first author Suzanne Hood, a doctoral student at Concordia.

Story of Lymphatic System Expands to Include Chapter on Valve Formation

A century after the valves that link the lymphatic and blood systems were first described, St. Jude Children's Research Hospital scientists have detailed how those valves form and identified a gene that is critical to the process.

The gene is *Prox1*. Earlier work led by Guillermo Oliver, Ph.D., a member of the St. Jude Department of Genetics, showed *Prox1* was essential for formation and maintenance of the entire lymphatic vasculature. The lymphatic vasculature is the network of vessels and ducts that help maintain the body's fluid balance and serves as a highway along which everything from cancer cells to disease-fighting immune components moves.

Oliver is senior author of the new study, which appeared in the October 15 edition of the scientific journal *Genes & Development*.

The new research suggests that *Prox1* is also essential for proper formation of the one-way valves that control movement of fluid and nutrients from the lymphatic system into the blood stream. Researchers found evidence that the Prox1 protein also has a role in formation of the venous valves.

"Understanding how valves form is crucial to efforts to develop treatments for valve defects that affect both children and adults," said the paper's first author, R. Sathish Srinivasan, Ph.D., a research associate in the St. Jude Department of Genetics. Those defects are linked to a variety of problems including lymphedema and deep vein thrombosis, which are blood clots that form deep in veins and have the potential for causing life-threatening complications. Lymphedema is the painful and sometimes disfiguring swelling that can occur when lymph flow is disrupted.

For more than a decade, the lymphatic system has been a focus of Oliver's laboratory. The laboratory's contributions through the years include evidence that leaky lymphatic vessels might contribute to obesity. Oliver and his colleagues also demonstrated how the lymphatic system forms from Prox1-producing cells destined to become lymphatic endothelial cells (LECs) when they leave the developing veins and migrate throughout the body.

The investigators also showed the *Coup-TFII* gene is essential to the process. The Coup-TFII protein binds to the promoter region of the *Prox1* gene. The binding switches on production of the Prox1 protein that is required to create and maintain the lymphatic system.

The newer research builds on that earlier work from Oliver's laboratory. The latest study focused on the lymphovenous valves. These valves are found at just two locations in the body, on either side of the chest just under the clavicle bone where the lymphatic vessels intersect with the subclavian and internal jugular veins.

Working in mice, investigators discovered that these lymphovenous valves form from a newly identified subtype of endothelial cell found in developing veins. Like the LECs that form the lymphatic system, the newly identified endothelial cells make Prox1. But while the LECs leave the veins and migrate throughout the body, these endothelial cells stay put to form the lymphovenous valves.

Researchers demonstrated the process requires two copies of the *Prox1* gene. That ensures adequate levels of the Coup-TFII-Prox1 complex and with it enough Prox1 to build and maintain the lymphatic system. Mice engineered to carry a single copy of *Prox1* either did not survive or were born without lymphovenous and venous valves.

"If you have only one copy of *Prox1* you are going to have a reduction in the Coup-TFII Prox1 complex and so a dramatic reduction in the number of cells available to build the lympha Bats, dolphins, and mole rats inspire advances in ultrasound technology

Sonar and ultrasound, which use sound as a navigational device and to paint accurate pictures of an environment, are the basis of countless technologies, including medical ultrasound machines and submarine navigation systems. But when it comes to more accurate sonar and ultrasound, animals' "biosonar" capabilities still have the human race beat. But not for long. In a new project that studies bats, dolphins, and mole rats, Prof. Nathan Intrator of Tel Aviv University's Blavatnik School of Computer Science, in collaboration with Brown University's Prof. Jim Simmons, is working to identify what gives biosonar the edge over human-made technologies.

Using a unique method for measuring how the animals interpret the returning signals, Prof. Intrator has determined that the key to these animals' success is superior, real-time data processing. "Animal 'echolocations' are done in fractions of milliseconds, at a resolution so high that a dolphin can see a tennis ball from approximately 260 feet away," he says, noting that the animals are able to process several pieces of information simultaneously.

Their research, which has been reported in the *Journal of the Acoustical Society of America* and presented at the 2010 and 2011 MLSP conferences, could lead to cutting-edge navigation systems and more accurate medical imaging.

Detecting "Shape" from Sound

Biosonar animals send ultrasonic sounds called "pings" into the environment. The shape of the returning signals, or echoes, determines how these animals "see" their surroundings, helping them to navigate or hunt for prey. In a matter of tens of milliseconds, the neurons in the animal's brain are capable of a full-scale analysis of their surroundings represented in three dimensions, with little energy consumption. Even with the aid of a supercomputer, which consumes thousands of times more energy, humans cannot produce such an accurate picture, Prof. Intrator says. With echolocation, a bat can tell the difference between a fly in motion or at rest, or determine which of two fruits is heavier by observing their movements in the wind.

Intrigued by the quality of the natural world's biosonar over its human-made equivalents, Profs. Intrator and Simmons set out to study how biosonar animals perform echo location so quickly and accurately. Using an electronic system, they altered the frequency and noise levels of the echo returned to the animal.

By manipulating the echo, the researchers could determine what factors of the returning signal reduced an animal's ability to correctly analyse the returns. This in turn led to a better understanding of how the returning echoes are represented and analysed in the animal's brain.

A More Accurate View of the Human Body

Prof. Intrator and his fellow researchers have created mathematical models, involving machine learning and signal processing, that improve man's ability to interpret the echoes. This will lead to more accurate echo localisation and better resilience to background noise.

Once researchers gather more information about animal interpretation of biosonar, they will be able to mimic this technology for better ultrasound and sonar systems, says Prof. Intrator. "Animals explore pings with multiple filters or receptive fields, and we have demonstrated that exploring each ping in multiple ways can lead to higher accuracy," he explains. "By understanding sonar animals, we can create a new family of ultrasound systems that will be able to explore our bodies with more accurate medical imaging."

This could provide a variety of benefits to the medical field, such as earlier detection of defects in embryos or non-invasive detection of cancer tumours. Unlike an MRI or CT machine, which are large, expensive to operate, and often use dangerous radiation, the new generation of ultrasound machines could be used in a doctor's office at a fraction of the cost. The research could also benefit military reconnaissance efforts both underwater and Researcher finds elderly lose ability to distinguish between odours

Scientists studying how the sense of smell changes as people age, found that olfactory sensory neurons in those 60 and over showed an unexpected response to odour that made it more difficult to distinguish specific smells, putting them at greater risk from dangerous chemicals and poor nutrition. "We found clear changes in olfactory sensory neuron responses to odours for those 60 and up," said Professor Diego Restrepo, Ph.D., director of the Centre for NeuroScience at the University of Colorado School of Medicine who led the researchers. "When we presented two different odours to the olfactory sensory neurons of younger people they responded to one or the other. The sensory neurons from the elderly responded to both. This would make it harder for the elderly to differentiate between them."

According to the study published in the latest issue of *Neurobiology of Aging*, those losing their sense of smell are at a higher risk of malnutrition since taste and smell are closely related, they may also be unable to detect spoiled food, leaking gas or toxic vapours.

Researchers looked at 440 subjects in two age groups those 45-years-old and younger and those 60 and over. Their olfactory sensory neurons (OSNs) were tested for their responses to two distinct odours as well as subsets of those odours.

Restrepo wanted to determine if age-related differences in the function of OSNs might contribute to an impairment of the sense of smell. In a collaboration with Monell Chemical Senses Centre in Philadelphia, researchers biopsied cells from both age groups.

"Whereas cells from younger donors were highly selective in the odorants to which they responded, cells from older donors were more likely to respond to multiple odour stimuli... suggesting a loss of specificity," the study said.

The scientists had expected to find less OSNs in older subjects and they thought the neurons would be less likely to respond to stimuli. In fact, they found as many neurons in the old as the young but those over 60 could not differentiate between two odours, they blended together.

Whiskers Marked Milestone in Evolution of Mammals from Reptiles

Research from the University of Sheffield comparing rats and mice with their distance relatives the marsupial, suggests that moveable whiskers were an important milestone in the evolution of mammals from reptiles. Using high-speed digital video recording and automatic tracking, the research team, which was led by Professor Tony Prescott from the University's Department of Psychology, have shed light on how rodents such as mice and rats move their whiskers back-and-forth at high speed and in varying ways to actively sense the environment around them in a behaviour known as whisking. Whisking allows mice or rats to accurately determine the position, shape and texture of objects, make rapid and accurate decisions about objects, and then use the information to build environmental maps.

When running in a straight line, rats and mice move their whiskers back-and-forth the same amount on both sides. However when turning, they bias their whisker movements in the direction of the turn, and when the whiskers on one side of the head contact an object, those on the opposite side sweep round to gather more information. These active sensing strategies boost the information gained by the whiskers helping the animals to better understand their world through touch.

In their latest research, the team have shown that whisking like that of rodents, using these active sensing strategies, is also seen in a small South American marsupial the grey short-tailed opossum. This animal has many similarities to an early mammal that would have lived more than 125 million years ago; that is, around the same time that the evolutionary lines leading to modern rodents and marsupials diverged.

This evidence suggests that some of the first mammals may also have whisked like a modern mouse or rat, and that the appearance of moveable whiskers was pivotal in the evolution of mammals from reptiles. The research is published in *Philosophical Transactions of the Royal Society B* on 12 November 2011 and will also be presented on the same day at the Society for Neuroscience conference.

The earliest mammals were nocturnal, and tree-living. In order to successfully move around and thrive in this challenging environment these animals needed to effectively integrate information from multiple senses sight, sound, smell, and touch. Facial whiskers provided mammals with a new tactile sense not available to reptiles that could help them to get around in the dark.

In addition to continuing to investigate the similarities and differences between rodents and marsupials, the team is also using these insights from biological whisker sensing to develop animal-like robots that can use artificial whiskers to navigate without vision. These robots could have applications in search-and-rescue, particularly in environments, such as disaster sites, where vision is compromised by smoke or dust.

Professor Tony Prescott said: "This latest research suggests that alongside becoming warm-blooded, giving birth to live young, and having an enlarged brain, the emergence of a new tactile sense based on moveable facial whiskers was an important step along the evolutionary path to modern mammals. Although humans no longer have moveable whiskers they were a critical feature of our early mammalian ancestors

Scientists Discover How to Design Drugs that could Target Particular Nerve Cells

The future of drug design lies in developing therapies that can target specific cellular processes without causing adverse reactions in other areas of the nervous system. Scientists at the Universities of Bristol and Liège in Belgium have discovered how to design drugs to target specific areas of the brain. The research, led by Professor Neil Marrion at Bristol's School of Physiology and Pharmacology and published in this week's *Proceedings of National Academy of Sciences (PNAS),* will enable the design of more effective drug compounds to enhance nerve activity in specific nerves.

The team has been working on a subtype of ion channel called SK channels. Ion channels are proteins that act as pores in a cell membrane and help control the excitability of nerves.

Rather like an electrical circuit, ion channels work by allowing the flow of 'charged' potassium, sodium and calcium ions to enter or exit cell membranes through a network of pores formed by the channels, a subtype of which is the SK channel family. The researchers have been using a natural toxin found in bee venom, called apamin, known for its ability to block different types of SK channel. SK channels enable a flow of potassium ions in and out of nerve cells that controls activity. The researchers have taken advantage of apamin being able to block one subtype of SK channel better than the others, to identify how three subtype SK channels [SK1-3] can be selectively blocked.

Neil Marrion, Professor of Neuroscience at the University, said: “The problem with developing drugs to target cellular processes has been that many cell types distributed throughout the body might all have the same ion channels. SK channels are also distributed throughout the brain, but it is becoming obvious that these channels might be made of more than one type of SK channel subunit. It is likely that different nerves have SK channels made from different subunits. This would mean that developing a drug to block a channel made of only one SK channel protein will not be therapeutically useful, but knowing that the channels are composed of multiple SK subunits will be the key.”

The study's findings have identified how SK channels are blocked by apamin and other ligands. Importantly, it shows how channels are folded to allow a drug to bind. This will enable drugs to be designed to block those SK channels that are made of more than one type of SK channel subunit, to target the symptoms of dementia and depression more effectively.

Vincent Seutin, one co-author of the paper, said: “Our study also shows a difference in the way apamin and nonpeptidic (potentially a useful drug) ligands interact with the channel. This may have important implications in terms of drug design.”

Chapter 6

Fish Flu: Genetics Approach

A research team at the National Institute of Standards and Technology (NIST) has provided the first look* at a genetic structure that may play a critical role in the reproduction of the infectious salmon anemia virus (ISAV), more commonly known as the "fish flu." A scourge in fish farms with a mortality rate as high as 90 percent, ISAV was recently found in wild salmon in the Pacific Northwest for the first time, threatening an already dwindling population and the vast food web it supports. While there is a vaccine for the virus, it must be administered by injection—a task that is both cumbersome and economically impractical for the aquaculture industry. A drug or vaccine that prevents the spread of the disease by interfering with the virus' ability to replicate its genetic code (contained in eight segments of ribonucleic acid or RNA) would be far more practical for fish farmers and marine biologists to deliver.

Robert Brinson, a NIST scientist working at the Hollings Marine Laboratory (HML) in Charleston, S.C., and NIST colleagues Andrea Szakal and John Marino working at the Institute for Bioscience and Biotechnology Research (IBBR) in Rockville, Md., knew from the scientific literature that the family of viruses that includes both the many types of influenza,the causes of yearly human flu outbreaks and infectious salmon anemia, form "panhandle" structures in their genomic RNA. In human influenza, these panhandles are known to interact with proteins that begin the process of copying and replicating the virus.

Hypothesizing that the fish flu virus might function the same way, Brinson and his colleagues used high-resolution nuclear magnetic resonance (NMR) spectroscopy and thermal melting methods to look at the genetic structure in the same region of the ISAV RNA. They found that the ISAV genome does appear to have a panhandle "motif" (the poetic term used by geneticists to define a discrete nucleotide sequence that functions independently of the rest of the genome and directs a specific biological function). The NIST work provides the first experimental evidence and characterisation of this panhandle motif that may function similarly in the fish virus as the "green light" for viral RNA replication processes.

"The next step," Brinson says, "is to investigate the relevant proteins and how they interface with the RNA. What molecular features drive the protein to recognise the RNA? How is it binding, and how is it interacting with the RNA?" Brinson says that this work may facilitate the development of new approaches for interfering with the replication of these viruses to mitigate the effects of ISAV on the aquaculture of salmon. It also demonstrates that the salmon virus can be used as an experimental model for understanding the replication machinery of human and other related influenza viruses.

The HML is a unique partnership of governmental and academic agencies including NIST, NOAA's National Ocean Service, the South Carolina Department of Natural Resources, the College of Charleston and the Medical University of South Carolina. The HML NMR facility focuses on the multi-institutional mission of metabolomics, natural products and structural biology. The IBBR is a University System of Maryland joint research enterprise created to enhance collaboration among the University of Maryland College Park, The University of Maryland Baltimore and NIST

ICRISAT-led Team Cracks Pigeonpea Genome

Once referred to as an "orphan crop" mainly grown by poor farmers, pigeonpea is now set to join the world's league of major food crops with the completion of its genome sequence. The completed genome sequence of pigeonpea is featured as an advance online publication on Nov. 6, 2011 on the website of the journal *Nature Biotechnology*, the highest ranked journal in the area of biotechnology. The paper provides an overview of the structure and function of the genes that define what makes a pigeonpea plant. It also reveals valuable clues on how the genomic sequence can be useful to crop improvement for sustainable food production particularly in the marginal environments of Asia and sub-Saharan Africa.

Years of genome analysis by a global research partnership led by the International Crops Research Institute for the Semi-Arid Tropics (ICRISAT) based in Hyderabad, India have resulted in the identification of 48,680 pigeonpea genes. A couple of hundred of these genes were found unique to the crop in terms of drought tolerance, an important trait that can be transferred to other similar legume crops like soybean, cowpea or common bean that belong to the same family.

In the fight against poverty and hunger amid the threat of climate change, highly nutritious, drought-tolerant crops are the best bets for smallholder farmers in marginal environments to survive and improve their livelihoods. Pigeonpea, grown on about 5 million hectares in Asia, sub-Saharan Africa and South-Central America, is a very important food legume for millions of the poor in the semi-arid regions of the world. Known as the "poor people's meat" because of its high protein content, it provides a well-balanced diet when accompanied with cereals.

"The mapping of the pigeonpea genome is a breakthrough that could not have come at a better time. Now that the world is faced with hunger and famine particularly in the Horn of Africa brought about by the worst drought of the decades, science-based, sustainable agricultural development solutions are vital in extricating vulnerable dryland communities out of poverty and hunger for good," says ICRISAT Director General William D. Dar. "Modern crop improvement technologies for smallholder farmer crops such as pigeonpea will be crucial to speed up the development of improved varieties that can provide high yields and improved livelihoods, and at the same time meet the challenges of marginal environments and the threat of climate change and scarce natural resources," adds Dar.

Rajeev Varshney, the lead scientist and coordinator for the pigeonpea genome sequencing project explains how this breakthrough will unlock pigeonpea's potential. "Having the pigeonpea genome sequence as a reference will significantly speed up and reduce the cost of screening the 'good genes' within the stored pigeonpea seed collections in genebanks like that of ICRISAT. This also means dramatically reducing the cost of developing new improved varieties for farmers," says Varshney.

"At the moment, in general, it can take 6-10 years to breed a new variety. With the use of this genome sequence data, in the future, we could be breeding a new variety in just about 3 years." he adds.

"The pigeonpea collaboration with ICRISAT is a milestone in the partnership between India and China, showcasing the excellent working dynamics and understanding among Indian and Chinese genomics scientists. I hope more partnerships like this will be established in the future, and I believe this will surely bring a significant difference to the whole world," says Professor Huanming Yang, Chairman of BGI-Shenzhen, the world's largest genomics institute and a key partner of this project. India is the largest producer of pigeonpea, but crop productivity in the country, as well as in sub-Saharan Africa, is only less than 1 ton per hectare. An improved understanding of the pigeonpea genome will have a major impact on improved crop productivity, tackling pests and disease constraints in production, and improved resistance to harsh environments and the future variable climate. Pigeonpea is the first "orphan crop," the first "non-industrial crop" and the second food legume (after soybean) with a completed genome sequence. It is also the first time that a Consultative Group on International Agricultural Research (CGIAR) supported Centre like ICRISAT has led the genome sequencing of a food crop.

The sequencing was accomplished by a global research partnership, the International Initiative for Pigeonpea Genomics (IIPG), led by ICRISAT with partners such as BGI Shenzhen (China), US research laboratories like University of Georgia, University of California-Davis, Cold Spring Harbor Laboratory, and National Centre for Genome Resources, European research institutes like the UK-based National University of Ireland Galway and also support from the CGIAR Generation Challenge Programme, US National Science Foundation and in-kind contribution from the collaborating research insti BGI contributes whole genome sequencing and bioinformatics expertise to potato genome research

BGI (previously known as the Beijing Genomics, Institute), the largest genomic organisation in the world, announced today that it was among the research organisations comprising the Potato Genome Sequencing Consortium (PGSC) that completed the genome sequence and analysis of the tuber crop potato, published as an Advance Online Publication in *Nature*. This study marks an important milestone in Potato (*Solanum tuberosum L.*) genome research, revealing new insights into the evolutionary history of the potato genome, causes of inbreeding depression, and potential mechanisms of tuber initiation and development. These insights will generate great interest among botanists and breeders worldwide and facilitate the genetic engineering of this vital crop. "BGI and its collaborators carried out the de novo sequencing, genome assembly and annotation of the potato genome, elucidating the evolutionary history of the genome and providing the foundation for biological analysis," stated Xun Xu, Vice President of Research and Development at BGI, and co-senior author of the report.

With the advanced genome sequencing capability of BGI, the homozygous doubled-monoploid DM1-3 516 R44 (DM) potato clone was sequenced by whole genome shotgun sequencing (WGS) approach with high coverage depths. The genome was then assembled by BGI's Short Oligonucleotide Analysis Package, SOAPdenovo, and 86% of the estimated total genome (844Mb) was assembled.

Researchers from BGI also aided in the potato gene annotation, of which 39,031 protein-coding genes were predicted. It is reported that at least two genome duplication events present in potato, indicating a paleopolyploid origin. Findings related with haplotype diversity, tuber biology and disease resistance are also reported. "We found important genetic basis related to inbreeding depression and tuber biology based on our bioinformatics analysis," said Xu Xun, who considers the potato as one of the most important plant genomes in light of growing food shortages due to global population growth and climate change.

BGI is the only genomic organisation with an extensive focus on agricultural genomics research. Its accomplishments include sequencing the rice genome, silkworm genome, cucumber genome, soybean genome and maize genome, among others. In January 2010, BGI launched the "1,000 Plant & Animal Reference Genomes Project" with the goal of generating reference genomes for 1,000 economically and scientifically important plant/animal species. " is being held in Shenzhen from July 10-12, 2011.

Reporting from the conference, Dr. Bicheng Yang, Director of Global Marketing at BGI stated, "Together with our collaborators, so far we have initiated 505 plant and animal genome projects, completed fine or draft genome maps for over 100 species and finished the sequencing of about 200 species. Many other genomes are undergoing active sequencing." She congratulated the PGSC consortium on its breakthrough in potato genome research and said BGI is proud to be a contributor to the project.

Last week, July 6, at BGI released SOAPdenovo 2, the latest update of BGI's SOAPdenovo package. SOAPdenovo 2 is reported to assemble genomes with improved efficiency and consistency, and can support a broader range of analyses. SOAPdenovo 2 can be downloaded at "BGI is making continuous efforts for the advancement of plant and animal genomes research. We will provide the most advanced infrastructure for our partners and look for more collaboration in this field." stated Dr. Yang.

Institute for Systems Biology uses Complete Genomics' Genome Sequencing Service to Verify Gene Responsible for Miller Syndrome

Complete Genomics Inc., a third-generation human genome sequencing company, today announced that the Institute for Systems Biology (ISB) employed Complete Genomics' human genome sequencing service to sequence a family quartet to determine the depth of genetic information possible in analysing a full family's sequence, and to verify the gene responsible for Miller syndrome, a rare craniofacial disorder. Results from this collaboration will be published online later today in the journal *Science*; the manuscript is titled "Analysis of Genetic Inheritance in a Family Quartet by Whole Genome Sequencing." "We are convinced that this new kind of analysis, family sequencing, will be a remarkably powerful scientific and medical tool in the future. ISB was delighted to work closely with the group at Complete Genomics to generate the data that enabled us to complete this study," said David Galas, Professor and Senior Vice President of ISB. ISB used Complete Genomics' service to sequence the genomes of a four-member nuclear family in which the two children suffer from Miller syndrome and ciliary dyskinesia, a lung disorder similar to cystic fibrosis, but neither parent is affected. The goal of this study was to identify rare genetic variations that could be responsible for Miller syndrome, and to estimate the intergenerational mutation rate. Complete Genomics sequenced the four genomes to a depth of 51x to 88x, and 85-92% of the bases in each genome were called. A comparative analysis of the four genomes yielded four genes consistent with recessive inheritance of rare variations. One of these genes, DHODH, was concurrently identified as a cause of Miller syndrome [Nat Genet. 2010 Jan;42(1):13-4]. Mutations in a second gene, DNAH5, have previously been shown to cause primary ciliary dyskinesia.

"When we established Complete Genomics, our goal was to provide large-scale complete human genome sequencing as a service that would enable our customers to make medically-relevant discoveries. We are delighted that ISB is already making breakthroughs of that caliber from its first study using our service," said Dr. Clifford Reid, chairman, president and CEO of Complete Genomics. "This is the type of positive disruptive influence that we want our technology to have on medical research."

Discovered Gene Causes Kabuki Syndrome

Using a new, rapid and less expensive DNA sequencing strategy, scientists have discovered genetic alterations that account for most cases of Kabuki syndrome, a rare disorder that causes multiple birth defects and mental retardation. Instead of sequencing the entire human genome, the new approach sequences just the exome, the 1-2 percent of the human genome that contains protein-coding genes. Kabuki syndrome, which has an estimated incidence of 1 in 32,000 births, was originally described by Japanese scientists in 1981. Patients with the disorder often have distinct facial features that resemble the make-up worn by actors of Kabuki, a Japanese theatrical form.

The work, published in today's advanced online edition of Nature Genetics, was carried out by scientists at the University of Washington in Seattle as part of a larger effort to use 'second generation' DNA sequencing technologies in new ways to identify genes for rare disorders. The project is funded by a $3.9 million American Recovery and Reinvestment Act grant from the National Human Genome Research Institute (NHGRI), part of the National Institutes of Health. "It's clear from this work that new DNA sequencing technologies are powerful and effective tools that scientists can use to accelerate the discovery of genes involved in rare diseases, an effort that previously was slow and costly," said NHGRI Director Eric D. Green, M.D., Ph.D. "The potential to rapidly identify gene mutations causing more than 6,000 rare diseases is an important step forward for researchers who are trying to understand the biology of these conditions and thereby improve strategies to care for patients they affect."

The University of Washington researchers sequenced the exomes of 10 unrelated individuals with Kabuki syndrome. Beginning with the premise that Kabuki syndrome is caused by alterations in just a single gene, the researchers compared the exomes of the 10 patients to the human genome sequence to look for differences shared by the affected individuals. Initially, the researchers identified none. The researchers then tested the hypothesis that Kabuki syndrome is more genetically heterogeneous than originally assumed, and that multiple genes could potentially cause the disorder.

Looking for novel variants in genes that were shared among subsets of the 10 patients' exomes, they found novel, matching variants that were shared in three genes in nine of the patient's exomes, shared in six genes in at least eight exomes, and shared in 16 genes among seven exomes. With no obvious way to rank these gene candidates, the researchers categorised each Kabuki case based on a subjective assessment of how well the patient matched the characteristics of Kabuki syndrome. They also looked in particular for variants in genes that led to a loss of function. The combined analysis pointed to gene called MLL2. The researchers found novel variants which lead to a loss of function in the MLL2 gene in the four highest ranked cases, and in three of the remaining six cases. The variants were nonsense and frameshift mutations. The nonsense mutation substituted a single nucleotide (a molecule that links to form DNA) in the gene code, while the frameshift mutation resulted in a four nucleotide deletion in the gene. In these cases, each nonsense and frameshift mutation resulted in the production of a shortened, nonfunctional protein. The MLL2 gene normally encodes a protein important in the regulation of chromatin, a protein that helps package the DNA in a compact form and allows the chromosome to fit in the cell nucleus. Changes in chromatin structure are associated with DNA replication and turning genes on or off. Once the researchers suspected the MLL2 gene, follow-up sequencing using the traditional and highly accurate Sanger DNA sequencing method, more capable of reliably detecting frameshifts, was used to identify additional MLL2 loss of function gene variants, which lead to a loss of function in two of the three remaining cases. Ultimately, the researchers were able to discover MLL2 mutations in nine of the 10 patients' exomes.

The findings were then validated using Sanger sequencing in 43 additional Kabuki syndrome cases. Novel MLL2 variants were found in 26 of the 43 cases. In the end, a total of 33 distinct MLL2 mutations were found in 35 of 53, or 66 percent, of patients with Kabuki syndrome. The researchers also discovered that in each of the 12 cases for which DNA from both parents was available, the MLL2 variants reflected new mutations within the affected individual's genome and were not present in either parent's genomes. "Our findings strongly suggest that alterations in the MLL2 gene are a major cause of Kabuki syndrome," said co-senior author Jay Shendure, M.D., Ph.D., assistant professor of genome sciences at the University of Washington School of Medicine. "It is clear that there may be additional genes in which variants cause Kabuki syndrome, as approximately one-third of cases did not have MLL2 mutations," said co-senior author Michael Bamshad, M.D., professor of pediatrics at the University of Washington School of Medicine. "To find these, it will be important to sequence the exomes of additional, well-characterised cases of Kabuki syndrome in which we do not find we don't see MLL2 mutations."

Recent Advances in Biology and Immunobiology of Eimeria Species and in Diagnosis and Control of Infection with These Coccidian Parasites of Poultry

Avian coccidiosis, an intestinal disease caused by protozoan parasites of the genus Eimeria, occurs worldwide. It is considered to be one of the most economically important diseases of domestic poultry. For many years, prophylactic use of anticoccidial feed additives has been the primary means of controlling coccidiosis in the broiler industry and has played a major role in the growth of this industry, which now can produce about 7.6 billion chickens annually. However, development of anticoccidial resistance has threatened the economic stability of the broiler industry.

Although there has been little effort by the pharmaceutical industry to develop new anticoccidials, the mounting problem of drug resistance of Eimeria species has prompted major research efforts to seek alternative means of control through increased knowledge of parasite biology, host response, and nutritional modulation. As a consequence, important advancements have been made, particularly in defining parasite antigens that have potential use in vaccines, defining the Eimeria genome, understanding the immunology of coccidial infections, and the practical applications of live vaccines. This review describes the progress in these areas, most of which has occurred within the past 10 to 15 years. Coccidiosis is recognised as the parasitic disease that has the greatest economic impact on poultry production. The annual worldwide cost is estimated at about $800 million (107), and that for the American broiler industry about $450 million. These estimates include the costs of prophylactic in-feed medication for broilers and broiler-breeders, alternative treatments (e.g., with amprolium) if the medications fail, and losses due to mortality, morbidity, and poor feed conversions of birds that survive outbreaks.

Life Cycle, Genetics, and Biochemistry

Avian Eimeria spp. have homoxenous life cycles, which have been well described (38). A web site that clearly details the biology of coccidia has also become available. In general, oocysts shed in feces undergo sporogony (a meiotic process) in the external environment, a process requiring oxygen, taking about 24 h. The sporulated oocysts contain four sporocysts, each containing two sporozoites. Upon ingestion by a suitable host, oocysts excyst within the intestinal lumen. This process is aided by trypsin, bile, and CO_2. The released sporozoites penetrate the villous epithelial cells. Sporozoites of some species (E. brunetti and E. praecox) develop within cells at the site of penetration. Sporozoites of other species (E. acervulina, E. maxima, E. necatrix, and E. tenella) are transported to other sites, for example the crypt epithelium, where they undergo development.

Within the host cells, sporozoites undergo asexual reproduction (schizogony or merogony) in which nuclear division is followed by cytoplasmic differentiation, resulting in merozoites that break free and penetrate other host cells. These may carry out several more merogonic generations. Sexual reproduction or gametogeny follows the last merogonic cycle. Merozoites enter host cells and develop into either male (microgamonts) or female (macrogamonts) forms. The microgamonts give rise to many microgametes, that exit, seek, and penetrate (fertilize) the macrogamonts that then develop into oocysts. Oocysts are shed with the feces. Prepatent periods may generally range from 4 to 5 days post-infection. Maximum oocyst output ranges from 6 to 9 days postinfection. The biochemical and genetic mechanisms that control the development of Eimeria spp. within host cells are not known. However, through study of precocious and drug-resistant lines of E. tenella (86, 87), two linkage groups associated with these traits have been identified and mapped to parasite chromosomes 1 and 2. This information may help identify other genetic loci involved in regulating the life cycle of E. tenella. Genetic segregation data from this study are available through the internet Other researchers have recently identified a gene (ets3a) whose expression is developmentally regulated and which may be important in controlling the life cycle of E. tenella. More genetic information on Eimeria and many other parasites can be found on the internet. Some of the websites include Eimeria.

The metabolism of Eimeria spp. has come under new scrutiny with the increase in our efforts to define protective antigens for use in vaccines and to target metabolic pathways for chemotherapy. For example, a mannitol cycle has been characterised in E. tenella (12, 80) that apparently provides mannitol as an energy source for sporulation. This pathway may be widespread within Eimeria spp. Newly characterised enzymes in E. tenella and E. maxima have been reported also, as well as a list of enzyme activities previously reported in avian Eimeria.

Species Determination and Diagnosis

There are seven valid species of chicken coccidia, E. acervulina, E. brunetti, E. maxima, E. mitis, E. necatrix, E. praecox, and E. tenella, each species developing in a particular location within the chick digestive tract. It is common to find at least six species (e.g., E. acervulina, E. maxima, E. tenella, E. brunetti, E. mitis, and E. praecox) in litter samples from a single flock during its first 6 weeks.

Five of the species, E. acervulina, E. brunetti, E. maxima, E. necatrix, and E. tenella, are well known and identifiable with relative ease, because they produce characteristic gross lesions. Their pathogenicities range from moderate to severe. On the other hand, E. praecox and E. mitis do not kill chickens or produce pathognomic lesions and often have been considered to be benign. However, experimental infections result in enteritis, diarrhea, and reduced feed efficiencies, indicating that these two species certainly can cause commercial losses and hence need to be controlled.

Generally, the five most pathogenic species listed above can be differentiated in the host on the basis of clinical signs, characteristic lesions at particular sites of infection in the chicken intestine, and consideration of the prepatent period, size of oocysts, and morphology of intracellular stages. However, the less pathogenic species such as E. mitis and E. praecox might be overlooked. A very useful handbook is available which describes standard diagnostic and testing procedures for Eimeria species and includes coloured photographs of gross lesions caused by the most commonly found species.

Shirley was the first to use a molecular biological approach to differentiate species on the basis of isoenzyme patterns of oocysts by starch block electrophoresis. Ellis and Bumstead were among the first to demonstrate that rRNA and rDNA probes could be used to identify individual species through characteristic restriction fragment patterns. Procunier et al. used a randomly amplified polymorphic DNA assay to differentiate E. acervulina and E. tenella and detect within-strain differences. Recombinant DNA techniques have been used to discriminate different strains of E. tenella and develop markers for precocious and drug-resistant strains, and PCR amplification of internal transcribed spacer region 1 from genomic DNA has been used to detect and differentiate six Eimeria species. Eight species (including E. hagani) are claimed to be differentiated using a two-step PCR procedure, and six Australian species have been characterised using a PCR-linked restriction fragment length polymorphism approach. These PCR methods should prove very useful for epidemiological surveys of avian coccidiosis.

Bibliography

Acharya, Manjushree: *Marine Biology*, International Scientific Pub, Delhi, 2011.

Allen, G. R. : *Freshwater Fishes of Australia*, T.F.H. Publications, New Jersey, 1989

Archana Prabhakar: *Fish Immunology and Biotechnology*, Swastik Publications, Delhi, 2010.

Arun Kumar: *Fish Diversity in the Selected Streams of Chakrata and Shiwalik Hills*, Zoological Survey of India, 2006.

Ausubel, F.M. : *Current Protocols in Molecular Biology,* New York: John Wiley and Sons, 1989.

Barman, R P : *Marine and Estuarine Fish Fauna of Orissa*, Zoological Survey of India, Delhi, 2007.

Barnes, N.: *Biology*, New York, Worth Publishers, 1989.

Barnum, Susan R.: *Biotechnology: An Introduction*, Belmont, Thomson/Brooks/Cole, 2005.

Barrington, E. J. W.: *Biochemistry of Primitive Deuterostomians*, London, Academic Press, 1974.

Bauer MW: *Biotechnology-the Making of a Global Controversy*, Cambridge, Cambridge University Press, 2002.

Bhargava, Gopal: *Marine Ecosystem*, Kalpaz, Delhi, 2001.

Bora, K.K. : *Agros Dictionary of Plant Physiology and Biochemistry,* Agrobios, Delhi, 2001.

Bose, Bandana and A. Hemantaranjan : *Developments in Physiology, Biochemistry and Molecular Biology of Plants,* New India Pub, Delhi, 2005.

Boyer , J.S.: *Measuring the Water Status of Plants and Soils,* Academic Press, N.Y., 1995.

Broach, J.R.: *The Molecular Biology of the Yeast,* Cold Spring Harbor, Cold Spring Harbor Laboratory, 1981.

Bruce M. : *Patten's Foundation of Embryology,* New York: McGraw-Hill, Inc., 1996.

Cahill, Lisa : *Genetics, Theology, and Ethics: An Interdisciplinary Conversation,* New York: Crossroad, 2005.

Chakraborty, Chiranjib : *Advances in Biochemistry and Biotechnology,* Daya, Delhi, 2005.

Chandrawati Jee and Shagufta: *Fish Biotechnology*, A.P.H. Pub, Delhi, 2010.

Chauhan, B.S. : *Principles of Biochemistry and Biophysics,* Laxmi Publications, Delhi, 2008.

Chhapgar, B F : *Marine Life in India,* Oxford University Press, Delhi, 2006.

Daphne C. Elliott: *Biochemistry and Molecular Biology*, Oxford University Press, Delhi, 2005.

Dar, Ghulam Hassan : *Soil Microbiology and Biochemistry,* New India Publishing Agency, Delhi, 2010.

Dayal, Rajeshwar: *Marine Biodiversity*, Surendra Pub, Delhi, 2011.

Dharamvir Hota: *Modern Biotechnology in Plant Breeding*, Gene Tech Books, Delhi, 2007.

Duckworth, W. L. H.: *Morphology and Anthropology : A Handbook for Students*, Cosmo, Delhi, 2006.

Dutta, Jyotirmoy: *Dawn of Modern Marine Science in India : Collection of Papers on Natural History*, The Asiatic Society, Delhi, 2001.

Farrand, John: *The Audubon Society Master Guide to Birding,* Alfred A. Knopf, New York. 1983

Featherly H. I.: *Taxonomic Terminology of the Higher Plants*, USA, Iowa State College Press, 1954.

Felix, S. : *Marine and Aquaculture Biotechnology*, Agrobios, Delhi, 2010.

Fletcher, H. O. : *Australian Museum Magazine,* Collecting in central and northern Australia, 1937.

Ganguly, Smriti : *Biochemistry of Biomolecules,* Pearl Books, Delhi, 2007.

Gotshall, D.W.: *Guide to Marine Invertebrates, Alaska to Baja California,* Sea Challengers. Monterey, California. 1994.

Hardy B.: *Biology and Agronomy of Forage Arachis*, Cali, International Centre for Tropical Agriculture, 1994.

Ivantsoff, W. : *Natural History of Dalhousie Springs*. W. F. South Australian Museum, Adelaide. 1989.

Jayasankar, P. and B. Anoop: *Identification of Marine Mammals of India*, Narendra Pub, Delhi, 2010.

Kamla Devi and D.V. Rao : *Poisonous and Venomous Fishes of Andaman Islands, Bay of Bengal,* Zoological Survey of India, 2003.

Kapoor, R.L. and M.L. Saini: *Plant Breeding and Crop Improvement*, CBS, Delhi, 1997.

Khanna, V K : *Objective Genetics, Biotechnology, Biochemistry and Forestry,* I.K. International Publishing House, Delhi, 2008.

Kuppuram, G. and K. Kumudamani: *Marine Archaeology : The Global Perspectives*, Sundeep, Delhi, 1996.

Lake, J. S. : *Freshwater Fishes and Rivers of Australia,* Thomas Nelson Australia Ltd, Melbourne, 1971.

Larry V. McIntire : *Biotechnology: Science, Engineering, and Ethical Challenges for the Twenty-first Century,* Washington, DC: Joseph Henry Press, 1996.

Madan Lal Bagdi: *Physiology, Biochemistry and Biotechnology*, Manglam Pub, Delhi, 2007.

Mahadevan, R.K. : *Fish Harvesting and Processing*, Campus Books International, Delhi, 2008.

Martin, K.C. : *Freshwater Fishes of the Northern Territory*. Northern Territory Museum of Arts and Sciences, Darwin. 1990.

Mitra, Abhijit and Kakoli Banerjee: *Marine Microbiology*, Narendra Pub, Delhi, 2004.

Mohanty, A.K. : *Entrepreneurship in Agriculture : Scopes and Opportunities*, Agrotech Pub, Delhi, 2011.

Muneesh Kainth: *Chordate Embryology*, Dominant, Delhi, 2003.

Narvekar, Raghunath : *Molecular Biochemistry : Principles and Practices,* Adhyayan Pub, Delhi, 2008.

Pandey, Ashok : *Biotechnology : Food Fermentation (Microbiology, Biochemistry and Technology)* , Educational Publishers, Delhi, 1999.

Pillai, N.G.K.: *Marine Fisheries and Mariculture in India*, Narendra, Delhi, 2011.

Pollard, D. A. *Freshwater Fishes of South-eastern Australia*. McDowell, R. M. Reed, Sydney. 1980.

Ramakrishnan, T.V. : *Marine and Offshore Engineering*, Gene Tech Books, Delhi, 2007.

Razdan, Dheeraj: *Encyclopaedia of Marine Insurance : Its Principles and Practices*, Cyber Tech Pub, Delhi, 2011.

Rekha R Gaonkar; Maria D C Rodrigues and R B Patil: *Fishery Management*, APH, Delhi, 2008.

Sahoo, Dinabandhu and Prem Chand Pandey: *Advances in Marine and Antarctic Science*, APH, Delhi, 2002.

Sangeeta Malvee: *Fish Genetics*, SBS Pub, Delhi, 2008.

Saradha, T. : *Biology and Effects of Endosulfan and Dimethoate on Marine Molluscs*, Nidhi, Delhi, 2004.

Selvamani, B.R. and R.K. Mahadevan: *Marine Capture Fisheries*, Campus Books, Delhi, 2008.

Sharma, R K and S P S Sangha : *Basic Techniques in Biochemistry and Molecular Biology,* I K International, Delhi, 2009.

Sheela Srivastava : *Genetics Embryology and Fishes,* Mangalam Pub Delhi, 2009.

Swarnim, K. : *A Textbook of Biochemistry and Microbiology,* Surendra Pub, Delhi, 2010.

Tomi Petr: *Fisheries in Irrigation Systems of Arid Asia*, Daya, Delhi, 2007.

Warren, E.B. and Herbert, R.A.: *Fishes of California and Western México. Pacific Marine Fishes,* T.F.H. Publications, Inc. Ltd. 1984.

Yadav, M : *Nutritional Biochemistry and Metabolism,* Arise Pub, Delhi, 2008.

Index

OOO